Chemistry Library
179
Advances in Polymer Science

QD
281
P6
F66
v. 179
CHEM

Editorial Board:
A. Abe • A.-C. Albertsson • R. Duncan • K. Dušek • W.H. de Jeu
J.F. Joanny • H.-H. Kausch • S. Kobyashi • K.-S. Lee • L. Leibler
T.E. Long • I. Manners • M. Möller • O. Nuyken • E. M. Terentjev
B. Voit • G. Wegner

ANAL

QD
281
P6
F66
v. 179
CHEM

Advances in Polymer Science

Recently Published and Forthcoming Volumes

**Intrinsic Molecular Mobility and
Toughness of Polymers II**
Volume Editor: Kausch H.H.
Vol. 188, 2005

**Intrinsic Molecular Mobility and
Toughness of Polymers I**
Volume Editor: Kausch H.H.
Vol. 187, 2005

Polysaccharides I
Structure, Characterisation and Use
Volume Editor: Heinze T.T.
Vol. 186, 2005

**Advanced Computer Simulation
Approaches for Soft Matter Science II**
Volume Editors: Holm C., Kremer K.
Vol. 185, 2005

Crosslinking in Materials Science
Vol. 184, 2005

Phase Behavior of Polymer Blends
Volume Editor: Freed K.
Vol. 183, 2005

Polymer Analysis · Polymer Theory
Vol. 182, 2005

**Interphases and Mesophases in Polymer
Crystallization II**
Volume Editor: Allegra G.
Vol. 181, 2005

**Interphases and Mesophases in Polymer
Crystallization I**
Volume Editor: Allegra G.
Vol. 180, 2005

**Inorganic Polymeric Nanocomposites
and Membranes**
Vol. 179, 2005

Polymeric and Inorganic Fibres
Vol. 178, 2005

Poly(arylene ethynylene)s
From Synthesis to Application
Volume Editor: Weder C.
Vol. 177, 2005

Metathesis Polymerization
Volume Editor: Buchmeiser M.R.
Vol. 176, 2005

Polymer Particles
Volume Editor: Okubo M.
Vol. 175, 2005

Neutron Spin Echo in Polymer Systems
Authors: Richter D.,
Monkenbusch M., Arbe A., Colmenero J.
Vol. 174, 2005

**Advanced Computer Simulation
Approaches for Soft Matter Sciences I**
Volume Editors: Holm C., Kremer K.
Vol. 173, 2005

Microlithography · Molecular Imprinting
Vol. 172, 2005

Polymer Synthesis
Vol. 171, 2004

**NMR · 3D Analysis ·
Photopolymerization**
Vol. 170, 2004

Long-Term Properties of Polyolefins
Volume Editor: Albertsson A.-C.
Vol. 169, 2004

Polymers and Light
Volume Editor: Lippert T.K.
Vol 168, 2004

New Synthetic Methods
Vol. 167, 2004

CD 8/11/05

Inorganic Polymeric Nanocomposites and Membranes

With contributions by
O. Becker · B. Boutevin · F. Guida-Pietrasanta · N. Hasegawa
M. Klapper · P. V. Kostoglodov · D. Likhatchev · K. Müllen
M. Kato · A. L. Rusanov · G. P. Simon · A. Usuki

 Springer

This series presents critical reviews of the present and future trends in polymer and biopolymer science including chemistry, physical chemistry, physics and material science. It is adressed to all scientists at universities and in industry who wish to keep abreast of advances in the topics covered.

As a rule, contributions are specially commissioned. The editors and publishers will, however, always be pleased to receive suggestions and supplementary information. Papers are accepted for "Advances in Polymer Science" in English.

In references Advances in Polymer Science is abbreviated *Adv Polym Sci* and is cited as a journal.

The electronic content of *Adv Polym Sci* may found at springerlink.com

Library of Congress Control Number: 20045921922

ISSN 0065-3195
ISBN-10 3-540-25325-4 **Springer Berlin Heidelberg New York**
ISBN-13 978-3-540-25325-9 **Springer Berlin Heidelberg New York**
DOI 10.1007/b101390

This work is subject to copyright. All rights are reserved, whether the whole or part of the material is concerned, specifically the rights of translation, reprinting, reuse of illustrations, recitation, broadcasting, reproduction on microfilm or in any other way, and storage in data banks. Duplication of this publication or parts thereof is permitted only under the provisions of the German Copyright Law of September 9, 1965, in its current version, and permission for use must always be obtained from Springer. Violations are liable for prosecution under the German Copyright Law.

Springer is a part of Springer Science+Business Media
springeronline.com
© Springer-Verlag Berlin Heidelberg 2005
Printed in Germany

The use of registered names, trademarks, etc. in this publication does not imply, even in the absence of a specific statement, that such names are exempt from the relevant protective laws and regulations and therefore free for general use.

Cover design: *Design & Production* GmbH, Heidelberg
Typesetting and Production: LE-TEX Jelonek, Schmidt & Vöckler GbR, Leipzig

Printed on acid-free paper 02/3141 YL – 5 4 3 2 1 0

Editorial Board

Prof. Akihiro Abe

Department of Industrial Chemistry
Tokyo Institute of Polytechnics
1583 Iiyama, Atsugi-shi 243-02, Japan
aabe@chem.t-kougei.ac.jp

Prof. A.-C. Albertsson

Department of Polymer Technology
The Royal Institute of Technology
S-10044 Stockholm, Sweden
aila@polymer.kth.se

Prof. Ruth Duncan

Welsh School of Pharmacy
Cardiff University
Redwood Building
King Edward VII Avenue
Cardiff CF 10 3XF
United Kingdom
duncan@cf.ac.uk

Prof. Karel Dušek

Institute of Macromolecular Chemistry
Academy of Sciences of the
Czech Republic
Heyrovský Sq. 2
16206 Prague 6, Czech Republic
dusek@imc.cas.cz

Prof. Dr. W. H. de Jeu

FOM-Institute AMOLF
Kruislaan 407
1098 SJ Amsterdam, The Netherlands
dejeu@amolf.nl
and

Dutch Polymer Institute
Eindhoven University of Technology
PO Box 513
5600 MB Eindhoven, The Netherlands

Prof. Jean-François Joanny

Physicochimie Curie
Institut Curie section recherche
26 rue d'Ulm
75248 Paris cedex 05, France
jean-francois.joanny@curie.fr

Prof. Hans-Henning Kausch

EPFL SB ISIC GGEC
J2 492 Bâtiment CH
Station 6
CH-1015 Lausanne, Switzerland
kausch.cully@bluewin.ch

Prof. S. Kobayashi

Department of Materials Chemistry
Graduate School of Engineering
Kyoto University
Kyoto 615-8510, Japan
kobayasi@mat.polym.kyoto-u.ac.jp

Prof. Kwang-Sup Lee

Department of Polymer Science &
Engineering
Hannam University
133 Ojung-Dong
Daejeon 306-791,Korea
kslee@mail.hannam.ac.kr

Prof. L. Leibler

Matière Molle et Chimie
Ecole Suprieure de Physique
et Chimie Industrielles (ESPCI)
10 rue Vauquelin
75231 Paris Cedex 05, France
ludwik.leibler@espci.fr

Prof. Timothy E. Long

Department of Chemistry
and Research Institute
Virginia Tech
2110 Hahn Hall (0344)
Blacksburg,VA 24061, USA
telong@vt.edu

Prof. Ian Manners

Department of Chemistry
University of Toronto
80 St. George St.
M5S 3H6 Ontario, Canada
imanners@chem.utoronto.ca

Prof. Dr. Martin Möller

Deutsches Wollforschungsinstitut
an der RWTH Aachen e.V.
Pauwelsstraße 8
52056 Aachen, Germany
moeller@dwi.rwth-aachen.de

Prof. Oskar Nuyken

Lehrstuhl für Makromolekulare Stoffe
TU München
Lichtenbergstr. 4
85747 Garching, Germany
oskar.nuyken@ch.tum.de

Dr. E. M. Terentjev

Cavendish Laboratory
Madingley Road
Cambridge CB 3 OHE
United Kingdom
emt1000@cam.ac.uk

Prof. Brigitte Voit

Institut für Polymerforschung Dresden
Hohe Straße 6
01069 Dresden, Germany
voit@ipfdd.de

Prof. Gerhard Wegner

Max-Planck-Institut
für Polymerforschung
Ackermannweg 10
Postfach 3148
55128 Mainz, Germany
wegner@mpip-mainz.mpg.de

Advances in Polymer Science
Also Available Electronically

For all customers who have a standing order to Advances in Polymer Science, we offer the electronic version via SpringerLink free of charge. Please contact your librarian who can receive a password or free access to the full articles by registering at:

springerlink.com

If you do not have a subscription, you can still view the tables of contents of the volumes and the abstract of each article by going to the SpringerLink Homepage, clicking on "Browse by Online Libraries", then "Chemical Sciences", and finally choose Advances in Polymer Science.

You will find information about the

– Editorial Board
– Aims and Scope
– Instructions for Authors
– Sample Contribution

at springeronline.com using the search function.

Contents

Polysilalkylene or Silarylene Siloxanes Said Hybrid Silicones
F. Guida-Pietrasanta, B. Boutevin 1

Epoxy Layered Silicate Nanocomposites
O. Becker, G.P. Simon 29

Proton-Exchanging Electrolyte Membranes
Based on Aromatic Condensation Polymers
A.L. Rusanov, D. Likhatchev, P.V. Kostoglodov, K. Müllen, M. Klapper . 83

Polymer-Clay Nanocomposites
A. Usuki, M. Kato 135

Author Index Volumes 101–179 197

Subject Index 217

Adv Polym Sci (2005) 179: 1–27
DOI 10.1007/b104479
© Springer-Verlag Berlin Heidelberg 2005
Published online: 6 June 2005

Polysilalkylene
or Silarylene Siloxanes Said Hybrid Silicones

F. Guida-Pietrasanta (✉) · B. Boutevin

Laboratoire de Chimie Macromoléculaire, UMR 5076 CNRS,
Ecole Nationale Supérieure de Chimie de Montpellier, 8 Rue de l'Ecole Normale,
34296 Montpellier Cedex 5, France
francine.guida-pietrasanta@enscm.fr, bernard.boutevin@enscm.fr

1	Introduction	2
2	Synthesis of "Hybrid" Silicones Starting from Bis-Silanol Monomers	4
2.1	From Bis-Silanol Monomers Obtained via an Organometallic Route	4
2.1.1	Aryl and/or Alkyl Backbone	4
2.1.2	Fluorinated Backbone	10
2.2	From Bis-Silanol Monomers Obtained Through Hydrosilylation	14
3	Synthesis of "Hybrid" Silicones Through Hydrosilylation of α, ω-Dienes. Hydrosilylation Polymerization	19
4	Conclusions	24
	References	25

Abstract This paper reviews different methods of synthesis of polysilalkylene or silary-lene siloxanes that are sometimes called "hybrid" silicones. This special type of silicone has been developed to avoid the drawback of the depolymerization of classical polysilox-anes in certain conditions and to obtain elastomers with enhanced thermal and fuel resistance properties. These silicones have been prepared through two main routes: the polycondensation of α, ω-bis silanol monomers (prepared either via an organometallic route or via hydrosilylation of α, ω-dienes) and the polyhydrosilylation of α, ω-dienes with dihydrodisiloxanes or oligosiloxanes.

Keywords Fluorinated polysiloxanes · Hydrosilylation · Polycarbosiloxanes · Polycondensation · Polysilalkylene siloxanes · Polysilarylene siloxanes

Abbreviations

BPA	bisphenol A
DMS	dimethylsiloxane
HSCTs	high speed civil transports
PDMS	polydimethylsiloxane
PI/PS	poly(imidesiloxanes)
Pt-DVTMDS	platinum-divinyl-1,3 tetramethyldisiloxane
PTFPMS	polytrifluoropropylmethylsiloxane

ScCO$_2$ supercritical carbon dioxide
TMG/CF$_3$CO$_2$H   tetramethylguanidine/trifluoroacetic acid
TMPS-DMS   tetramethyl-p-silphenylenesiloxane-dimethylsiloxane

1
Introduction

Classical polysiloxanes – $[(R)(R')SiO]_n$ – have been extensively studied and some of them were already commercialized as early as the 1940s. Their various properties allowed applications in such various fields as aeronautics, biomedical, cosmetics, waterproof surface treatment, sealants, unmolding agents, etc. What is particularly interesting with silicones is the great flexibility of their backbones, due to the OSiO chainings, which induces a very low glass transition temperature (T_g), and also their low surface tension which makes them hydrophobic. These two properties account for their wide range of applications despite their high cost.

They also exhibit a rather good thermal stability, but in certain conditions (in acid or base medium or at high temperature) they may depolymerize due to chain scission of some SiOSi moieties through a six centers mechanism [1] (cf. Fig. 1), and give rise to cycles and shorter linear chains.

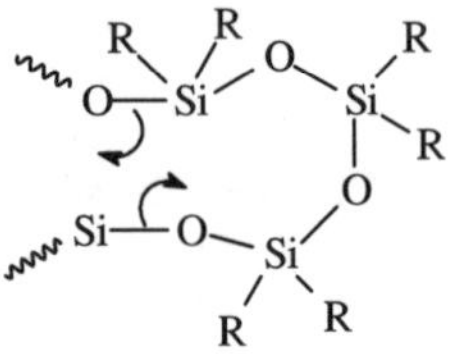

Fig. 1

This intramolecular cycloreversion may occur from at least 4 SiO bonds [2].

So, several researchers have shown interest in another type of polysiloxane: polysilalkylenesiloxanes or hybrid silicones alternating SiO and SiC bonds in their backbones and having the following general formula:

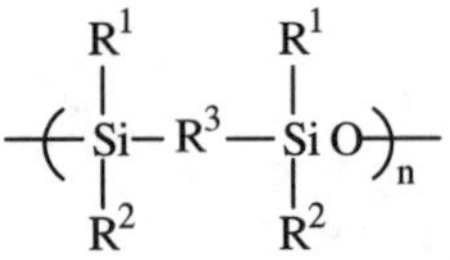

Fig. 2

where R^3 may be an alkyl, aryl, alkyl aryl or fluoroalkyl chain.

Several synthetic routes have been described in the literature to obtain these polysiloxanes. They will be examined hereafter.

One of the first examples of hybrid silicone was published in 1955 [3] by Sommer and Ansul, who reported the obtention of hybrid "paraffin-siloxanes" containing the 1,6-disilahexane group which was synthesized as follows:

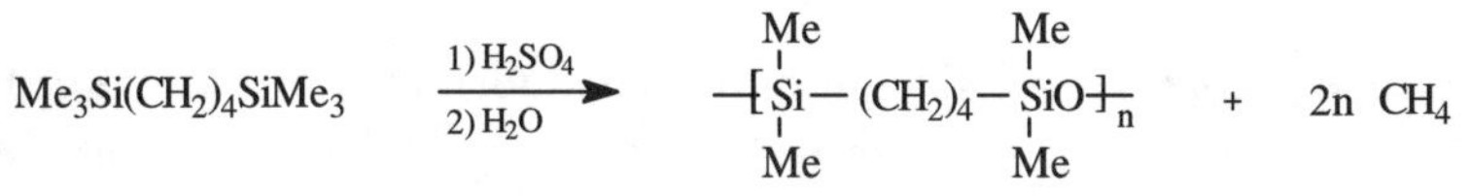

$$Me_3Si(CH_2)_4SiMe_3 \xrightarrow[\text{2) } H_2O]{\text{1) } H_2SO_4} \quad -\!\!\left[\!\begin{array}{c} Me \\ | \\ Si \\ | \\ Me \end{array}\!-(CH_2)_4-\!\!\begin{array}{c} Me \\ | \\ SiO \\ | \\ Me \end{array}\!\right]_n \quad + \quad 2n\ CH_4$$

Scheme 1

This hybrid silicone was presented as a compound having an intermediate structure between linear methylpolysiloxanes and paraffin hydrocarbons.

Then, during the years 1960–1970 many other examples of hybrid silicones were described, particularly silphenylene-siloxanes that are hybrid silicones containing phenyl groups in the backbone of the siloxane chain, and also fluorinated hybrid silicones with or without aromatic groups in the backbone or as side chains.

These silicones are generally obtained using two main pathways:

1. From bis-silanol monomers, themselves prepared either via an organo-metallic route or via hydrosilylation of α, ω-dienes. The bis-silanol monomers are then polymerized to give hybrid homopolymers or condensed with difunctional silanes to give copolymers (cf. Scheme 2).

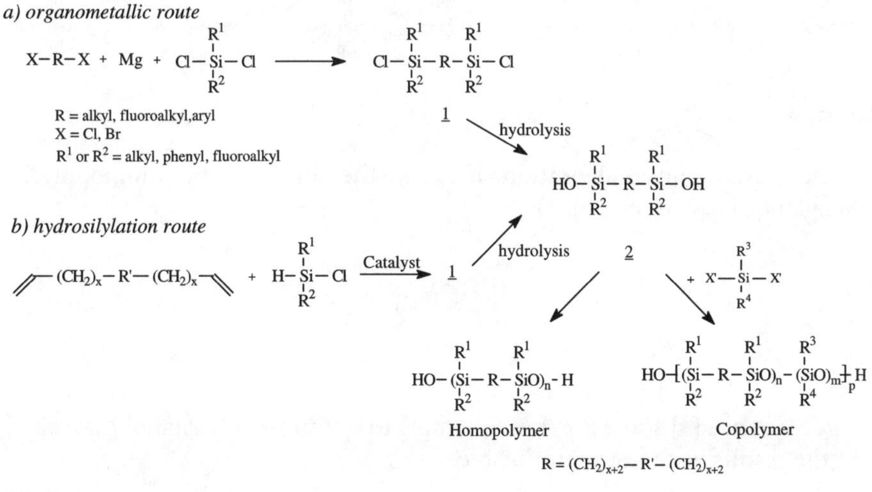

a) organometallic route

$$X-R-X + Mg + \begin{array}{c} R^1 \\ | \\ Cl-Si-Cl \\ | \\ R^2 \end{array} \longrightarrow \begin{array}{c} R^1 \quad\quad R^1 \\ | \quad\quad\ | \\ Cl-Si-R-Si-Cl \\ | \quad\quad\ | \\ R^2 \quad\quad R^2 \end{array}$$

R = alkyl, fluoroalkyl, aryl
X = Cl, Br
R^1 or R^2 = alkyl, phenyl, fluoroalkyl

1 $\searrow$ hydrolysis

$$\begin{array}{c} R^1 \quad\quad R^1 \\ | \quad\quad\ | \\ HO-Si-R-Si-OH \\ | \quad\quad\ | \\ R^2 \quad\quad R^2 \end{array}$$

b) hydrosilylation route

$$\diagup\!\!\!\!-(CH_2)_x-R'-(CH_2)_x-\!\!\!\diagdown \quad + \quad \begin{array}{c} R^1 \\ | \\ H-Si-Cl \\ | \\ R^2 \end{array} \xrightarrow{\text{Catalyst}} \mathbf{1} \quad \diagup \text{hydrolysis} \quad \mathbf{2} \quad + \quad \begin{array}{c} R^3 \\ | \\ X'-Si-X' \\ | \\ R^4 \end{array}$$

$$\begin{array}{c} R^1 \quad\quad R^1 \\ | \quad\quad\ | \\ HO-(Si-R-SiO)_n-H \\ | \quad\quad\ | \\ R^2 \quad\quad R^2 \end{array} \qquad\qquad \begin{array}{c} R^1 \quad\ R^1 \quad\ R^3 \\ | \quad\ | \quad\ | \\ HO-[(Si-R-SiO)_n-(SiO)_m]_p H \\ | \quad\ | \quad\ | \\ R^2 \quad\ R^2 \quad\ R^4 \end{array}$$

Homopolymer   Copolymer

$$R = (CH_2)_{x+2}-R'-(CH_2)_{x+2}$$

Scheme 2

2. Through polyaddition of α, ω-dienes with α, ω-dihydro di or oligosiloxanes, in other words by polyhydrosilylation (cf. Scheme 3).

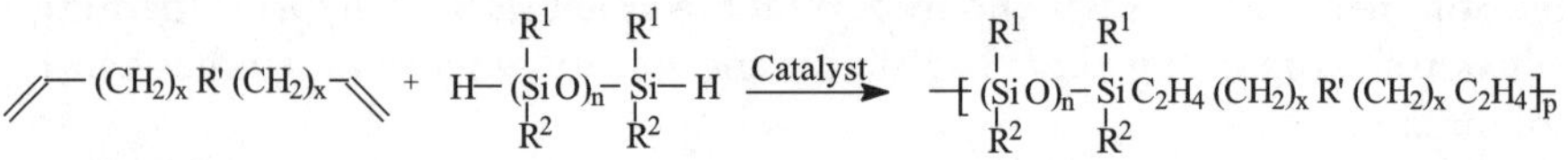

$$\text{//---(CH}_2)_x \text{ R' (CH}_2)_x \text{---} \quad + \quad \text{H---} \overset{\overset{\displaystyle R^1}{|}}{\underset{\underset{\displaystyle R^2}{|}}{(Si\,O)_n}} \text{---} \overset{\overset{\displaystyle R^1}{|}}{\underset{\underset{\displaystyle R^2}{|}}{Si}} \text{---H} \quad \xrightarrow{\text{Catalyst}} \quad \text{---}\overset{\overset{\displaystyle R^1}{|}}{\underset{\underset{\displaystyle R^2}{|}}{(Si\,O)_n}} \text{---}\overset{\overset{\displaystyle R^1}{|}}{\underset{\underset{\displaystyle R^2}{|}}{Si}} C_2H_4 \text{(CH}_2)_x \text{ R' (CH}_2)_x \text{ C}_2H_4\text{---}_p$$

Scheme 3

This review concerns silalkylene siloxanes fluorinated or non fluorinated, aromatic or nonaromatic, but we have voluntarily excluded polysilanes, i.e. polymers that contain silicon but without any SiOSi bonds.

2
Synthesis of Hybrid Silicones Starting from Bis-Silanol Monomers

2.1
From Bis-Silanol Monomers Obtained via an Organometallic Route

2.1.1
Aryl and/or Alkyl Backbone

One of the first hybrid bis-silanols that was used in the synthesis of hybrid silicones, and reported by Merker and Scott in 1964 [4], was bishydroxy(tetramethyl-p-silphenylene siloxane) $\underline{1}$. It was obtained via a magnesium route according to Scheme 4:

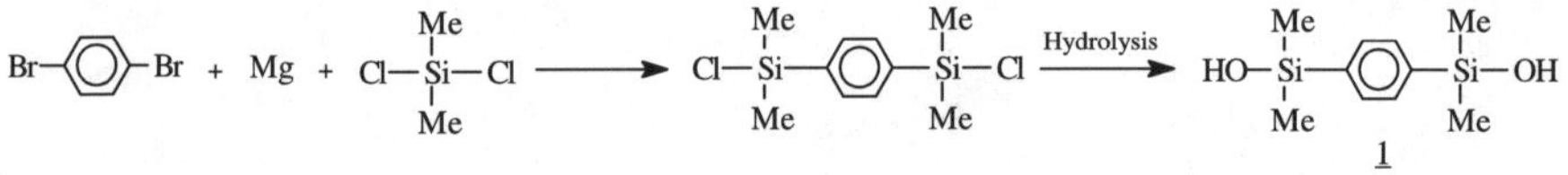

$$\text{Br---}\langle\bigcirc\rangle\text{---Br} + \text{Mg} + \text{Cl---}\overset{\overset{\displaystyle Me}{|}}{\underset{\underset{\displaystyle Me}{|}}{Si}}\text{---Cl} \longrightarrow \text{Cl---}\overset{\overset{\displaystyle Me}{|}}{\underset{\underset{\displaystyle Me}{|}}{Si}}\text{---}\langle\bigcirc\rangle\text{---}\overset{\overset{\displaystyle Me}{|}}{\underset{\underset{\displaystyle Me}{|}}{Si}}\text{---Cl} \xrightarrow{\text{Hydrolysis}} \text{HO---}\overset{\overset{\displaystyle Me}{|}}{\underset{\underset{\displaystyle Me}{|}}{Si}}\text{---}\langle\bigcirc\rangle\text{---}\overset{\overset{\displaystyle Me}{|}}{\underset{\underset{\displaystyle Me}{|}}{Si}}\text{---OH}$$

$$\underline{1}$$

Scheme 4

Then, after polycondensation, it led to the corresponding homopolysilphenylene siloxane (cf. Fig. 3):

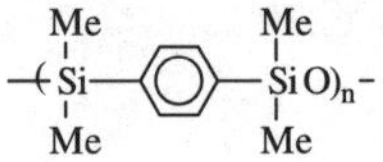

$$\text{---}\overset{\overset{\displaystyle Me}{|}}{\underset{\underset{\displaystyle Me}{|}}{Si}}\text{---}\langle\bigcirc\rangle\text{---}\overset{\overset{\displaystyle Me}{|}}{\underset{\underset{\displaystyle Me}{|}}{Si\,O}}_n\text{---}$$

Fig. 3

In reference [4] some previous attempts to synthesize bis-silanol $\underline{1}$ are cited but the results were not reproducible.

The hybrid homosilphenylenepolysiloxane presents a better thermal stability than polydimethylsiloxane (PDMS). It is solid (melting point = 148 °C instead of $-$ 40 °C for PDMS).

Bis-silanol $\underline{1}$ has been used in different syntheses of random, alternated or block copolymers. Merker et al. [5] described random and block copolymers of the following structure:

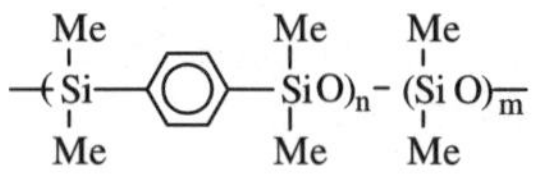

Fig. 4

These polymers were elastic at temperatures above their melting points (which may be up to 148 °C depending on the amount of oxysilphenylene component).

Alternating copolymers were supposed to exhibit the lower crystallinity and the higher thermal stability, but the authors were not able to obtain such polymers as their synthesis method (condensation of $\underline{1}$ with dimethyldichlorosilane) did not lead to alternance.

One year later, Curry and Byrd [6] obtained alternating copolymers by condensing diol $\underline{1}$ with diaminosilanes (cf. Scheme 5):

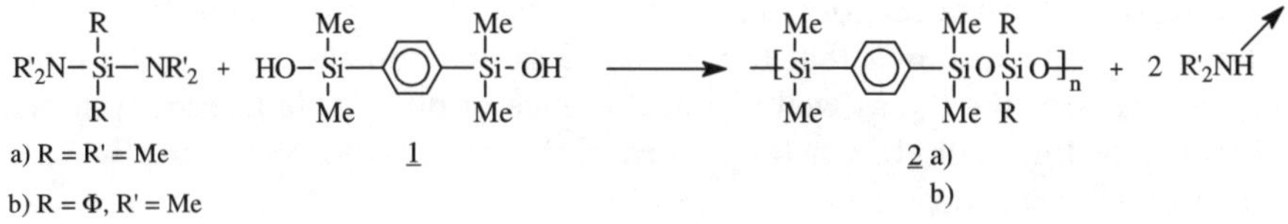

Scheme 5

This same reaction has been reproduced some years later by Burks et al. [7] and amorphous copolymers $\underline{2a}$ and $\underline{2b}$ were prepared, and studied as thermostable elastomers for the aeronautic industry. Copolymer $\underline{2a}$ or poly[1,4-bis(oxydimethylsilyl)benzene dimethylsilane] exhibited a glass transition temperature $T_g = -$ 63 °C and a very good stability at high temperature. Copolymer $\underline{2b}$ or poly[1,4-bis(oxydimethylsilyl)benzene diphenylsilane] exhibited a $T_g = 0$ °C and a higher stability at high temperature.

They were crosslinked at room temperature with Si(OEt)$_4$ and dibutyltin diacetate to give thermostable elastomers.

Since the beginning of the 1980s and during the 1990s, Dvornic and Lenz and their co-workers have published numerous articles on the synthesis of silarylene siloxanes and the study of their thermal properties [8–18]:

The synthesis was achieved according to Scheme 6:

$$HO-\underset{R^2}{\overset{R^1}{Si}}-\bigcirc-\underset{R^2}{\overset{R^1}{Si}}-OH \ + \ n \ X-\underset{R^4}{\overset{R^3}{Si}}-X \ \longrightarrow \ \underset{R^2}{\overset{R^1}{+Si}}-\bigcirc-\underset{R^2}{\overset{R^1}{Si}}-O-\underset{R^4}{\overset{R^3}{(Si\,O)}}_x\!\!\!\!+_n \ + \ 2n \ HX$$

Scheme 6

In reference [15], the authors compare four different ways to obtain these hybrid silarylene-siloxane copolymers:

1. $X = Cl$
2. $X = OCOCH_3$
3. $X = NMe_2$
4. $X = N(\Phi)CON <$ (ureido)

They conclude that the copolymers with higher molecular weights are obtained with ureidosilanes, while those with the lower molecular weights are obtained with chloro and acetoxy silanes, because in both these cases, degradation side reactions occur with the acids formed (HCl and CH_3COOH).

In the various publications [8–18], the nature of R^1 to R^4 were different: methyl, ethyl, vinyl, alkyl, phenyl, cyanoethyl, cyanopropyl, hydrogen and, more recently, fluoroalkyl [17].

The relations between the nature of the polymers and the glass transition temperatures have been studied [16], as well as their thermal stability [13, 14]. The authors have shown that the presence of an aromatic unit in the main chain increases the T_g, as well as the presence of bulky side groups (phenyl, cyanoalkyl, fluoroalkyl). On the contrary, the presence of vinyl or allyl side groups decreases the T_g.

Concerning thermal stability, the best resistance to thermal degradation is obtained with exactly alternating copolymers ($x = 1$ in Scheme 6).

In nitrogen, resistance to pure thermal degradation decreases depending on the type of the side groups R^3 and R^4 (when $R^1 = R^2 = CH_3$) in the following order:

$$CH = CH_2 > C_6H_5 > CH_3 > H > C_2H_4CF_3 > C_2H_4C_6F_{13}$$

So, the highest stability is observed with the vinyl group.

The synthesis and the properties of silphenylene-siloxanes have been summarized in a chapter of a monograph on silicon polymers [18]:

- DSC (Differential Scanning Calorimetry) measurements showed that the T_g increased when the size of the side groups increased.
- Thermal stability of these polymers is very high: in TGA (Thermal Gravimetric Analysis), they show decomposition not until 480 to 545 °C.

- The average molecular weights of the polymers range from 70 000 to 340 000.

More recently, in 1998 and 1999, McKnight et al. [19–21] reported some vinyl-substituted silphenylene siloxane copolymers with exactly alternating structures and varying vinyl content that were synthesized through disilanol diaminosilane polycondensation, as follows:

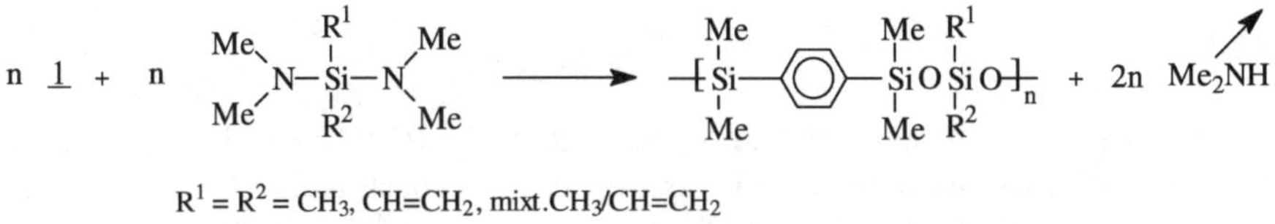

$$n \; \underline{1} \; + \; n \quad \begin{array}{c} Me \\ Me \end{array} \hspace{-0.3em} N \hspace{-0.3em} - \hspace{-0.3em} \overset{\overset{R^1}{|}}{\underset{\underset{R^2}{|}}{Si}} \hspace{-0.3em} - \hspace{-0.3em} N \hspace{-0.3em} \begin{array}{c} Me \\ Me \end{array} \longrightarrow \quad \begin{array}{c} Me \\ | \\ Si \\ | \\ Me \end{array} \hspace{-0.3em} - \hspace{-0.3em} \bigcirc \hspace{-0.3em} - \hspace{-0.3em} \overset{\overset{Me}{|}}{\underset{\underset{Me}{|}}{Si}} O \overset{\overset{R^1}{|}}{\underset{\underset{R^2}{|}}{Si}} O \hspace{-0.5em} \Big]_n \; + \; 2n \; Me_2NH$$

$$R^1 = R^2 = CH_3, \; CH=CH_2, \; mixt.CH_3/CH=CH_2$$

Scheme 7

The copolymers were described as thermally stable, high-temperature elastomers.

It was said that "they had low T_gs (ranging from − 26 to − 86 °C) and exhibited the highest degree of thermal and oxidation stability that has been observed so far for any elastomers". Additionally they were supposed to be promising candidates for potential applications as flame-retardant elastomers, one of the critical needs in many industrial branches such as the aircraft and automotive industry.

A few years earlier, in 1991, Williams et al. [22] had performed the structural analysis of poly(tetramethyl-p-silphenylene siloxane)-poly(dimethylsiloxane) copolymers (TMPS-DMS copolymers) by ^{29}Si NMR. These copolymers were obtained by the condensation of bis-hydroxy(tetramethyl-p-silphenylene siloxane) $\underline{1}$ with α, ω-dihydroxy polydimethyl oligosiloxanes, in the presence of a guanidinium catalyst (cf. Scheme 8):

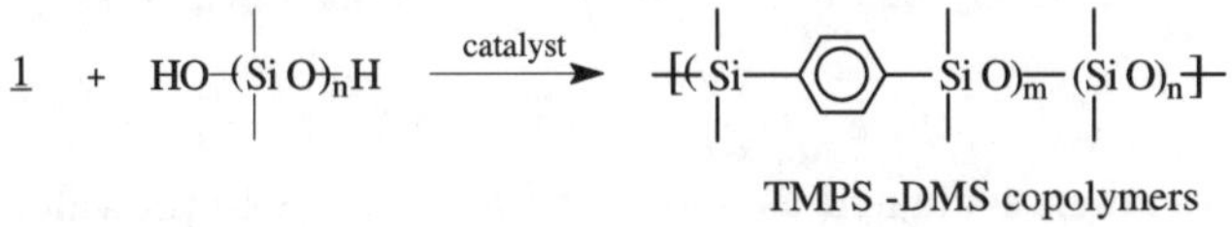

$$\underline{1} \; + \; HO \hspace{-0.3em}-\hspace{-0.3em}(Si\,O)_n H \xrightarrow{\text{catalyst}} \begin{array}{c} | \\ Si \\ | \end{array} \hspace{-0.3em}-\hspace{-0.3em} \bigcirc \hspace{-0.3em}-\hspace{-0.3em} \begin{array}{c} | \\ Si\,O \\ | \end{array} \Big]_m (Si\,O)_n \Big]$$

TMPS -DMS copolymers

Scheme 8

This NMR analysis is particularly useful as the block TMPS-DMS copolymers exhibit a wide range of properties depending upon the composition and average sequence lengths of the soft dimethylsiloxane segments and the hard crystalline silphenylene blocks.

In the years 1988 and 1989, in our laboratory [23, 24] the same bis-hydroxy (tetramethyl-p-silphenylene siloxane) $\underline{1}$ had been used in polycondensation with chlorosilanes fluorinated or nonfluorinated, type $Cl_2Si(Me)R^i$ with $R^i = H$, $CH = CH_2$, R_F and $R_F = C_3H_6OC_2H_4C_nF_{2n+1}$, $C_2H_4C_6F_5$, $C_3H_6OCF_2CFHCF_3$, $C_2H_4SC_2H_4C_nF_{2n+1}$, $C_3H_6SC_3H_6OC_2H_4C_nF_{2n+1}$ and silicones with the following general formula were obtained:

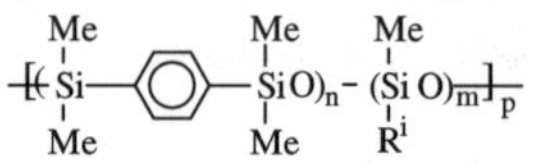

Fig. 5

Silicones containing, at the same time, $R^i = R_F$, $R^i = H$ and R^i = vinyl, are fluorinated silicones with low viscosities, easily crosslinkable by addition of Pt catalyst and that give access to "pumpable" fluorinated silicones.

Later, in 1997, we also described a hybrid silalkylene (C_6H_{12}) polysiloxane obtained by polycondensation of the corresponding hybrid bisilanol bearing methyl and phenyl pendant groups and showed that it also exhibited a good thermal stability [25]. Its $T_g = -52\,°C$ was higher than that of PDMS, but its degradation temperature in nitrogen was about 100 °C higher than for PDMS and was also higher in air.

Stern et al. [26] had published, in 1987, an article where they studied the structure-permeability relations of various silicon polymers and which gave, among others, the T_g of several hybrid silicones $-[(Me)_2Si - R - Si(Me)_2O]_x -$, where $R = -C_2H_4 -, -C_6H_{12} -, -C_8H_{16} - (T_g$s around $-90\,°C)$, $R = m\text{-}C_6H_4 - (T_g = -48\,°C)$ and $R = p\text{-}C_6H_4 - (T_g = -18\,°C)$, but nothing was said about their synthesis.

In fact, in 1997, the synthesis of poly(tetramethyl-m-silphenylene siloxane) was reported by Mark et al. [27] as follows:

Scheme 9

The T_g was then evaluated as $-52\,°C$ which is close to the value of $-48\,°C$ given by Stern et al. and no melting temperature was detected, contrary to the equivalent p-silphenylene polymer. TGA measurements revealed very good high temperature properties with the onset temperatures for degradation being 415 °C under nitrogen and 495 °C in air.

Finally, silarylene-siloxane-diacetylene polymers were reported by Homrighausen and Keller in 2000 [28], as precursors to high temperature elastomers. They were obtained as follows:

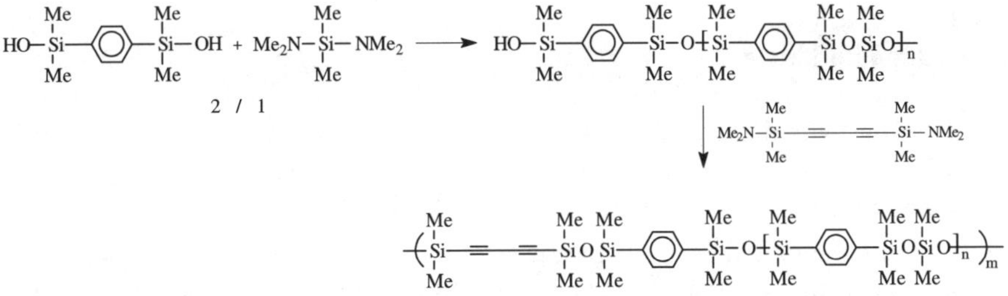

Scheme 10

Depending upon diacetylene content, the linear polymers can be transformed (via thermolysis) to either highly crosslinked plastics or slightly crosslinked elastomers. The crosslinked polymers degrade thermally above 425 °C under inert conditions.

As a variant of this first method using Grignard reagents to prepare hybrid silicones, it may be cited a very recently published synthesis of poly(siloxylene-ethylene-phenylene-ethylene)s by reaction of a bis-chlorosiloxane with the bismagnesium derivative of a diethynyl compound [29, 30] according to the following scheme:

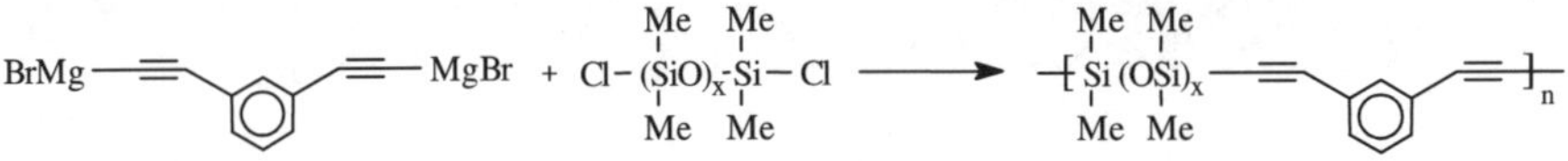

Scheme 11

These compounds are said to be useful for composites with good heat resistance.

The recent synthesis of silicon-containing fluorene polymers through the carbon-silicon coupling between fluorenyl Grignard reagents and dichlorosilanes may also be cited [31] (cf. Scheme 12).

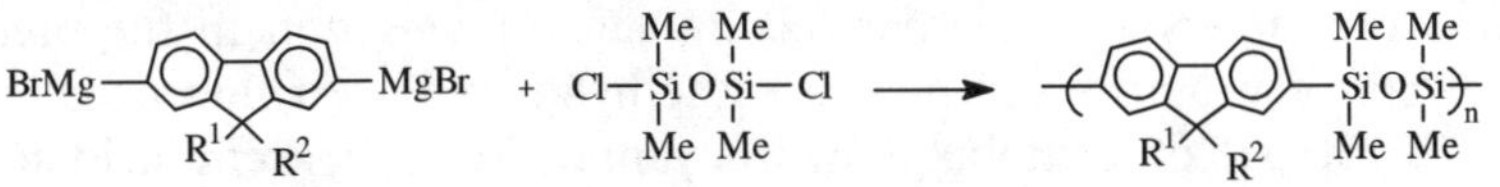

Scheme 12

Novel polymers have thus been prepared and their optical (UV-vis photoluminescence) and thermal properties have been studied.

2.1.2
Fluorinated Backbone

Concerning hybrid silicones fluorinated in the main chain, that are prepared from fluorinated hybrid bis-silanols obtained via a Grignard route, several examples may be cited:

- a patent deposited in 1970 by researchers from Dow Corning Corp. [32] describes the preparation of bis-silylfluoro-aromatic compounds and derivated polymers. The monomer diols, synthesized through Grignard reactions are of the type shown in Figs. 6 and 7:

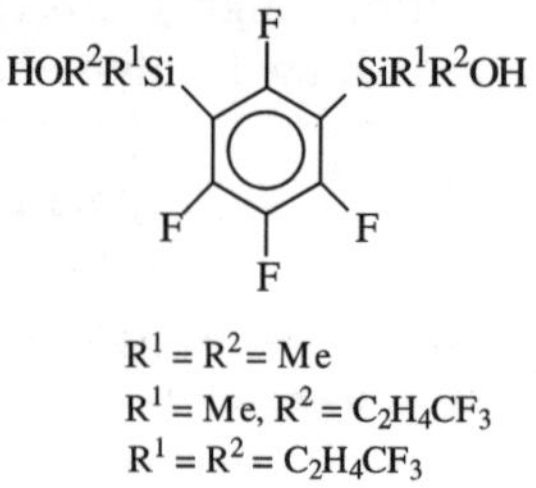

$$R^1 = R^2 = Me$$
$$R^1 = Me, R^2 = C_2H_4CF_3$$
$$R^1 = R^2 = C_2H_4CF_3$$

Fig. 6

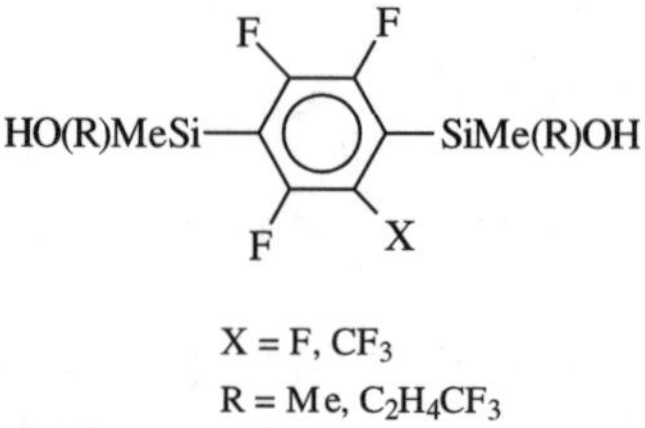

$$X = F, CF_3$$
$$R = Me, C_2H_4CF_3$$

Fig. 7

These monomers are polymerized by autocondensation in the presence of catalysts such as the complex tetramethylguanidine/trifluoroacetic acid (TMG/CF_3CO_2H) or tertiobutyl hydroxyamine/trifluoroacetic acid to give hybrid homopolymers (cf. Fig. 8):

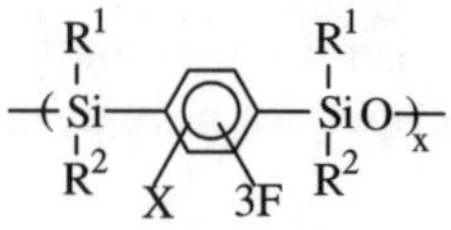

Fig. 8

After addition of charges, these polymers lead to elastomers that are stable at high temperature and have applications as sealant materials.

The diols monomers may also be co-hydrolysed with other siloxanes to give copolymers such as, for example Figs. 9, 10 and 11:

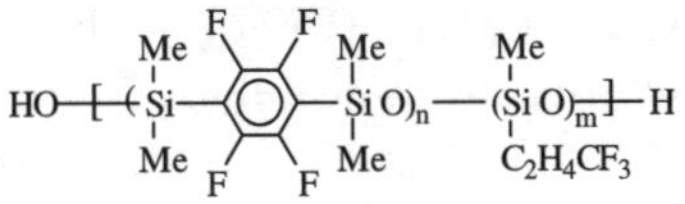

$$\text{HO} \left[\overset{\overset{\displaystyle Me}{|}}{\underset{\underset{\displaystyle Me}{|}}{Si}} - \underset{F \; F}{\overset{F \; F}{\bigcirc}} - \overset{\overset{\displaystyle Me}{|}}{\underset{\underset{\displaystyle Me}{|}}{Si}} O \right]_n - \left(\overset{\overset{\displaystyle Me}{|}}{\underset{\underset{\displaystyle C_2H_4CF_3}{|}}{Si}} O \right)_m H$$

Fig. 9

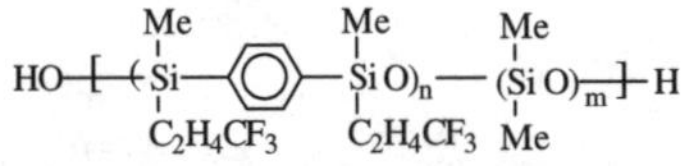

$$\text{HO} \left[\overset{\overset{\displaystyle Me}{|}}{\underset{\underset{\displaystyle C_2H_4CF_3}{|}}{Si}} - \bigcirc - \overset{\overset{\displaystyle Me}{|}}{\underset{\underset{\displaystyle C_2H_4CF_3}{|}}{Si}} O \right]_n - \left(\overset{\overset{\displaystyle Me}{|}}{\underset{\underset{\displaystyle Me}{|}}{Si}} O \right)_m H$$

Fig. 10

$$\text{HO} \left[\overset{\overset{\displaystyle Et}{|}}{\underset{\underset{\displaystyle Et}{|}}{Si}} - \underset{3F \; Br}{\bigcirc} - \overset{\overset{\displaystyle Me}{|}}{\underset{\underset{\displaystyle Me}{|}}{Si}} O \right]_n - \left(\overset{\overset{\displaystyle Me}{|}}{\underset{\underset{\displaystyle \Phi}{|}}{Si}} O \right)_m H$$

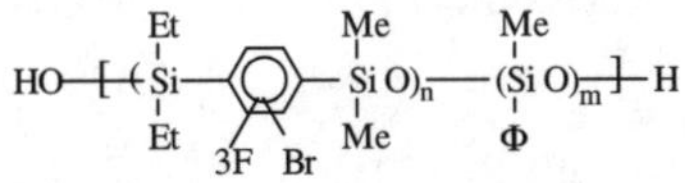

Fig. 11

- In parallel, another patent also deposited by Dow Corning Corp. [33] described the synthesis of silylfluoroaromatic homopolymers (cf. Scheme 13):

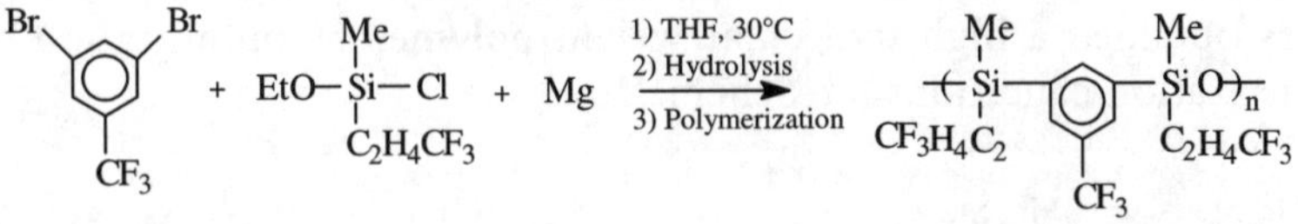

$$\underset{CF_3}{\overset{Br \quad Br}{\bigcirc}} + \overset{\overset{Me}{|}}{\underset{\underset{C_2H_4CF_3}{|}}{EtO-Si-Cl}} + Mg \quad \xrightarrow[\substack{3) \text{ Polymerization}}]{\substack{1) \text{ THF, 30°C} \\ 2) \text{ Hydrolysis}}} \quad \left[\overset{\overset{Me}{|}}{\underset{\underset{CF_3H_4C_2}{|}}{Si}} - \underset{CF_3}{\bigcirc} - \overset{\overset{Me}{|}}{\underset{\underset{C_2H_4CF_3}{|}}{Si}} O \right]_n$$

Scheme 13

- At the same time, Critchley et al. [34] published the synthesis of perfluoroalkylene organopolysiloxanes, still obtained from a monomer diol, that had been prepared by a Grignard route (cf. Scheme 14):

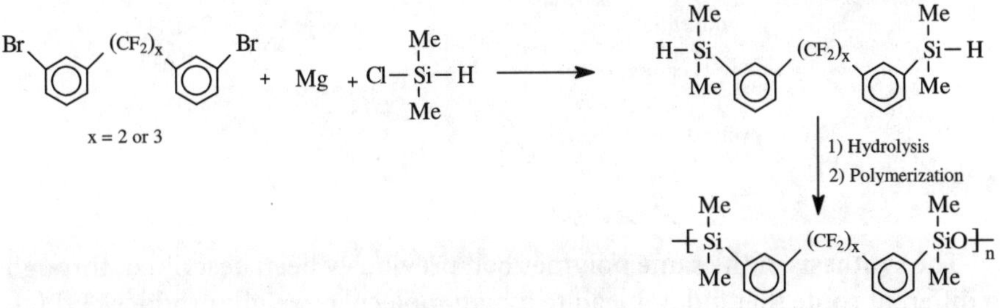

$$\overset{Br}{\bigcirc} (CF_2)_x \overset{Br}{\bigcirc} + Mg + \overset{\overset{Me}{|}}{\underset{\underset{Me}{|}}{Cl-Si-H}} \longrightarrow \quad \overset{\overset{Me}{|}}{\underset{\underset{Me}{|}}{H-Si}} \bigcirc (CF_2)_x \bigcirc \overset{\overset{Me}{|}}{\underset{\underset{Me}{|}}{Si-H}}$$

x = 2 or 3

$$\Bigg\downarrow \substack{1) \text{ Hydrolysis} \\ 2) \text{ Polymerization}}$$

$$\left[\overset{\overset{Me}{|}}{\underset{\underset{Me}{|}}{Si}} \bigcirc (CF_2)_x \bigcirc \overset{\overset{Me}{|}}{\underset{\underset{Me}{|}}{SiO}} \right]_n$$

Scheme 14

The study of the thermal degradation of these same hybrid silicones [35] was achieved in comparison to the classical polydimethyl and polytrifluo-ropropylmethyl siloxanes, and the authors showed that the introduction of perfluoroalkylene segments $- C_6H_4 - (CF_2)_x - C_6H_4 -$ into the main chain of the polysiloxane increased the thermal stability both under inert and oxidative atmosphere.

The same type of silphenylene siloxane polymers containing perfluoroalkyl groups in the main chain, was described by Patterson et al. [36, 37]. The starting diol monomers were also obtained via a Grignard route (cf. Scheme 15).

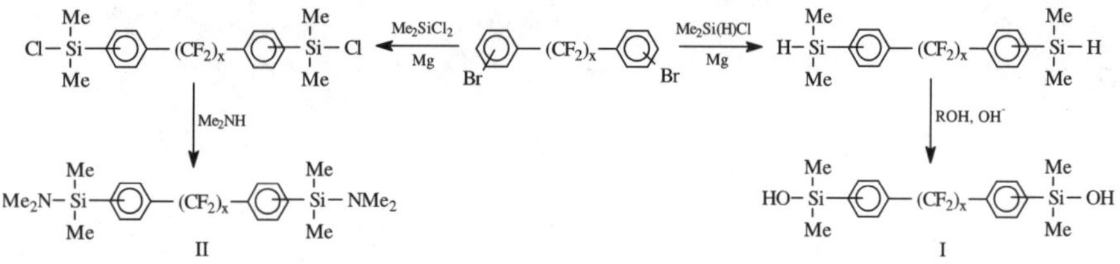

Scheme 15

Condensation of I and II led to hybrid silicon homopolymers that gave thermostable elastomers, after crosslinking.

The preparation of a fluorinated polysiloxane elastomer with silyl benzene moieties (called FASIL) was described by Loughran and Griffin [38]. The authors obtained a high molecular weight polymer by optimization of the polymerization conditions (cf. Scheme 16):

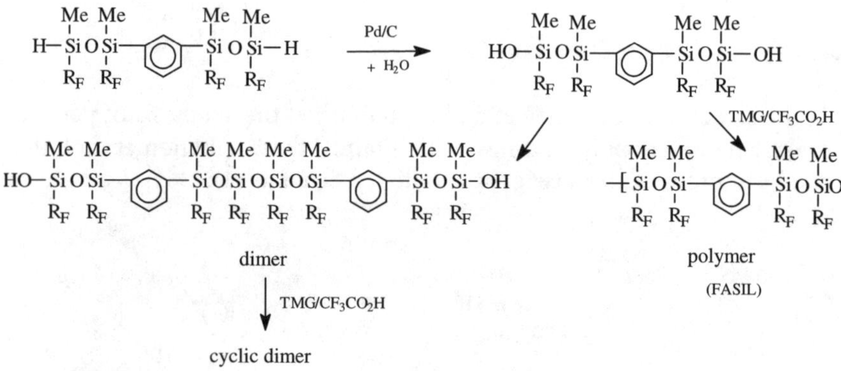

Scheme 16

The synthesis of the same polymer had previously been described, through a different route that did not lead to a high molecular weight product [39] (cf. Scheme 17):

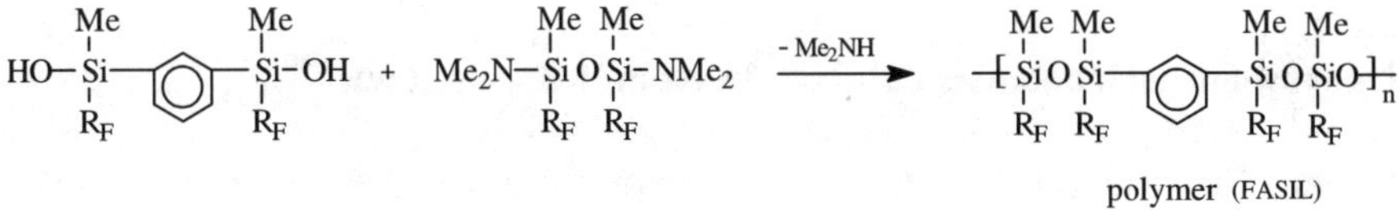

polymer (FASIL)

Scheme 17

Recently, Rizzo and Harris reported the synthesis and thermal properties of fluorosilicones containing perfluorocyclobutane rings [40] that can be considered as a particular kind of hybrid fluorinated silicones. Their work was directed towards "developing elastomers that could lead to high temperature fuel tank sealants that can be used at higher temperatures than the commercially available fluorosilicones." Actually, after base (KOH or NaH)-catalyzed self-condensation of the disilanol monomer, they obtained high molecular weight homopolymers (M_n ranging from $19\,000$ to $300\,000\,\text{g mol}^{-1}$) exhibiting very good thermal properties. The synthesis of the homopolymers was performed as follows:

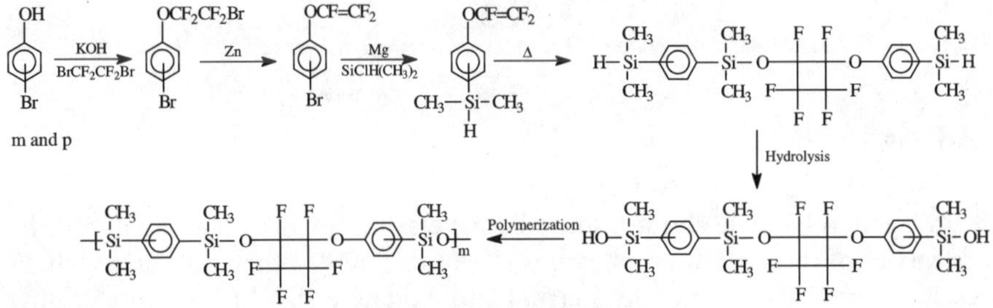

Scheme 18

The α, ω-bishydroxy homopolymers were also copolymerized with an α, ω-silanol terminated 3,3,3-trifluoropropyl methyl siloxane oligomer (classical fluorosilicone) to give copolymers with varying compositions. The T_gs of the copolymers ranging from -60 to $-1\,°\text{C}$, increased as the amount of perfluorocyclobutane-containing silphenylene repeat units increased. The TGA analysis showed that when the copolymers contained more than 20% of this repeat unit, they displayed less weight loss at elevated temperature than a classical fluorosilicone homopolymer. After crosslinking (using a peroxide) of a copolymer containing about 30 wt.% of the perfluorocyclobutane-containing repeating unit, the crosslinked network displayed a volume swell of under 40% in isooctane, similar to a crosslinked fluorosilicone.

2.2
From Bis-Silanol Monomers Obtained Through Hydrosilylation

During the year 1970, several articles were published by Kim et al. [41–46] about the synthesis and the properties of fluorinated hybrid silicone homopolymers and copolymers. These polymers were obtained by hydrosilylation of α, ω-dienes with chlorohydrogenosilanes, and the obtained bischlorosilanes were then hydrolysed into bis-silanols and polymerized or copolycondensed ($R^i = R^1$ or R^2 or R^3 or R^4, Z = alkyl, alkyl ether, fluoroalkyl, fluoroether, etc.) (cf. Scheme 19).

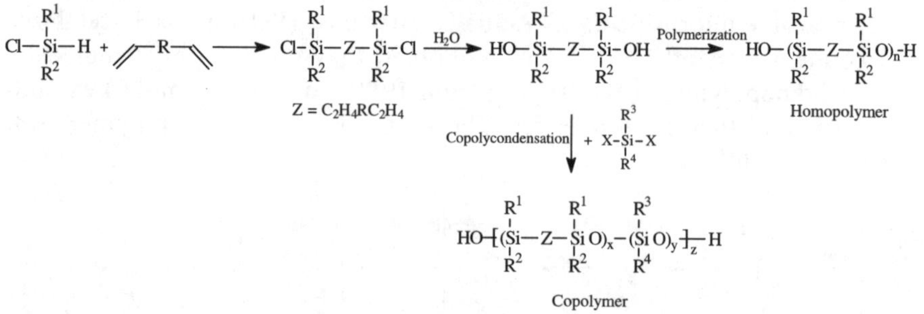

Scheme 19

In a general article about fluorosilicone elastomers [41], Kim analyzed the properties of classical fluorosilicones – $[(R)(R_F)SiO]_n$ – that are: "an excellent resistance to solvents, a good thermal and oxidative stability, an outstanding flexibility at low temperature." He concluded that fluorosilicones are superior to fluorocarbon elastomers, but they were not very good at high temperatures (above 450 °C). Conventional polydimethylsiloxanes, and classical fluorosilicones, present the drawback to give reversion or depolymerization at high temperature, which deteriorates the physical properties.

So, in order to obtain polymers that are resistant to reversion (or depolymerization) at high temperature, Kim decided to consider the synthesis of polymers of the type of Fig. 12:

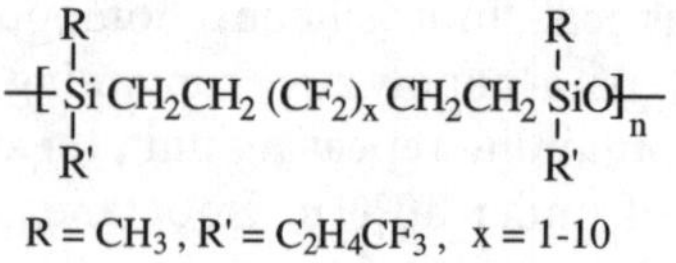

$$R = CH_3, R' = C_2H_4CF_3, \quad x = 1\text{-}10$$

Fig. 12

He recognized, then, that these types of compounds would be less flexible than classical silicones, at low temperature and thus would exhibit a higher T_g. Later, Kim et al. introduced a fluoroether segment Z into the homopolymers (cf. Scheme 19) and they showed that the thermal and oxidative stabilities of these new homopolymers were comparable to those of polymers as in Fig. 12, while their flexibility at low temperature was better, i.e. their T_g was lower [42] .They have synthesized numerous hybrid fluorosilicon homopolymers with $Z = CH_2CH_2RCH_2CH_2$ being fluoroalkyl or fluoroether (cf. Fig. 13):

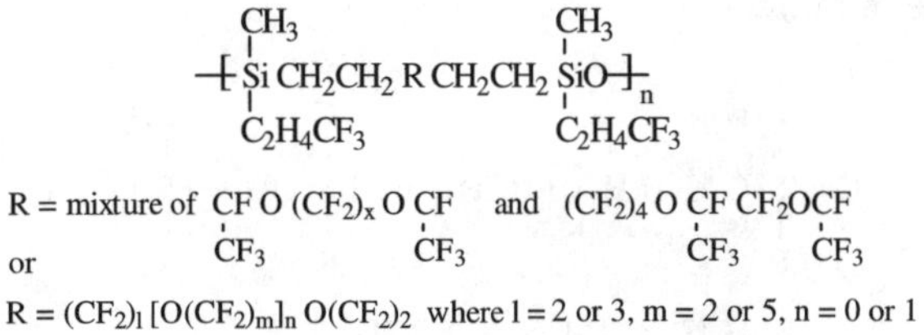

$$
\begin{array}{cc}
CH_3 & CH_3 \\
\mid & \mid \\
\text{--}\!\!\left[\text{Si } CH_2CH_2 \text{ R } CH_2CH_2 \text{ Si O}\right]\!\!\text{--}_n \\
\mid & \mid \\
C_2H_4CF_3 & C_2H_4CF_3
\end{array}
$$

$R =$ mixture of $CF O (CF_2)_x O CF$ and $(CF_2)_4 O CF CF_2OCF$
$\qquad\qquad\qquad\quad \underset{CF_3}{\mid}\qquad \underset{CF_3}{\mid}\qquad\qquad \underset{CF_3}{\mid}\quad \underset{CF_3}{\mid}$
or

$R = (CF_2)_l [O(CF_2)_m]_n O(CF_2)_2$ where $l = 2$ or 3, $m = 2$ or 5, $n = 0$ or 1

Fig. 13

Then, they considered fluorinated hybrid copolymers (cf. Scheme 20). These copolymers were prepared by condensation of hybrid bis-silanol monomers and dichloro or diacetamido silanes, in the presence of a monofunctional silane as the chain stopper, according to the following scheme:

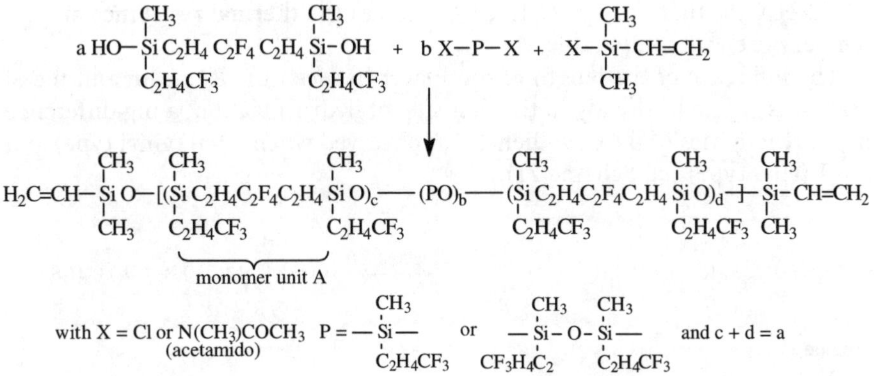

$$
\begin{array}{c}
CH_3 \qquad\qquad CH_3 \qquad\qquad\qquad\qquad\qquad CH_3 \\
\mid \qquad\qquad\quad \mid \qquad\qquad\qquad\qquad\qquad\qquad \mid \\
a\ HO\text{--}Si\ C_2H_4\ C_2F_4\ C_2H_4\ Si\text{--}OH \ + \ b\ X\text{--}P\text{--}X \ + \ X\text{--}Si\text{--}CH\!=\!CH_2 \\
\mid \qquad\qquad\qquad\qquad \mid \qquad\qquad\qquad\qquad\qquad\qquad \mid \\
C_2H_4CF_3 \qquad\qquad C_2H_4CF_3 \qquad\qquad\qquad\qquad CH_3
\end{array}
$$

$$\downarrow$$

$$
\begin{array}{c}
CH_3 \quad CH_3 \qquad\qquad CH_3 \qquad\qquad\qquad CH_3 \qquad\qquad CH_3 \quad CH_3 \\
\mid \qquad \mid \qquad\qquad\quad \mid \qquad\qquad\qquad\qquad \mid \qquad\qquad\quad \mid \qquad\quad \mid \\
H_2C\!=\!CH\text{--}Si\ O\text{--}[(Si\ C_2H_4C_2F_4C_2H_4\ Si\ O)_c\text{---}(PO)_b\text{---}(Si\ C_2H_4C_2F_4C_2H_4\ Si\ O)_d\text{--}]\text{--}Si\text{--}CH\!=\!CH_2 \\
\mid \qquad\quad \mid \qquad\qquad\qquad \mid \qquad\qquad\qquad\qquad \mid \qquad\qquad\quad \mid \qquad\quad \mid \\
CH_3 \quad C_2H_4CF_3 \qquad\qquad C_2H_4CF_3 \qquad\qquad\qquad C_2H_4CF_3 \qquad\quad C_2H_4CF_3 \quad CH_3
\end{array}
$$

$\underbrace{\qquad\qquad\qquad\qquad\qquad}_{\text{monomer unit A}}$

$$
\begin{array}{c}
\qquad\qquad\qquad\qquad CH_3 \qquad\qquad CH_3 \quad CH_3 \\
\qquad\qquad\qquad\qquad\qquad \mid \qquad\qquad\quad \mid \qquad\quad \mid \\
\text{with } X = Cl \text{ or } N(CH_3)COCH_3 \ \ P = \text{--}Si\text{--} \quad \text{or} \quad \text{--}Si\text{-}O\text{-}Si\text{--} \quad \text{and } c + d = a \\
\qquad\qquad\qquad (\text{acetamido}) \qquad\qquad\quad \mid \qquad\qquad\quad \mid \qquad\quad \mid \\
\qquad\qquad\qquad\qquad\qquad C_2H_4CF_3 \qquad CF_3H_4C_2 \quad C_2H_4CF_3
\end{array}
$$

Scheme 20

For $X = Cl$, they obtained random copolymers and for $X =$ acetamido, they obtained alternated copolymers $(AB)_n$ or $(ABA)_n$ depending on the nature of P [46], the monomer unit B being $- (CH_3)(C_2H_4CF_3)SiO -$.

A comparative study of the thermal properties and of the glass transition temperatures of the $(A)_n$ and $(B)_n$ homopolymers and of the $(AB)_n$ random and alternated copolymers and $(BAB)_n$ alternated copolymers has been achieved and showed the influence of the structure of the polymer.

Random copolymers may lead to depolymerization like $(B)_n$ homopoly-mers. On the contrary, alternated copolymers present a much better resis-tance to reversion. Copolymers exhibit a lower T_g (of 10 to 20 °C) than that of the hybrid homopolymer $(A)_n$. Thermogravimetric analyses of random and alternated copolymers show that they are more stable than each homopoly-mer $(A)_n$ or $(B)_n$.

More recently, in our laboratory, different homopolymers and copolymers comparable to those of Kim were synthesized [47–50] and products such as in Fig. 14 were obtained:

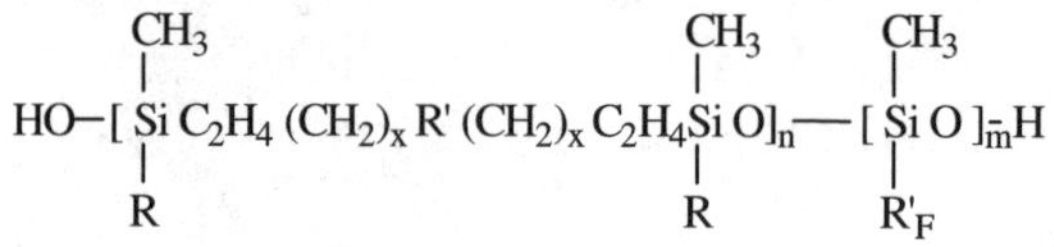

$$HO-[\underset{\underset{R}{|}}{\overset{\overset{CH_3}{|}}{Si}}\ C_2H_4\ (CH_2)_x\ R'\ (CH_2)_x\ C_2H_4\underset{\underset{R}{|}}{\overset{\overset{CH_3}{|}}{Si}}\ O]_n-[\underset{\underset{R'_F}{|}}{\overset{\overset{CH_3}{|}}{Si}}\ O]_{\overline{m}}H$$

$R = CH_3$, $C_2H_4CF_3$, $C_2H_4C_4F_9$

$R' = C_6F_{12}$, $CF(CF_3)CF_2$ C_4F_8 $CF_2CF(CF_3)$

$R'_F = C_2H_4C_6F_{13}$

x = 0 or 1, m = 0 or m

Fig. 14

It was shown that when the side chain R is fluorinated, the longer the fluorinated chain, the better the thermal resistance. The T_g was lower for $R = C_2H_4C_4F_9$ than for $R = C_2H_4CF_3$, whereas the thermal resistance at high temperature was comparable.

The influence of the length of the spacer between the R_F chain and the Si atom was studied. Already in the first step of hydrosilylation, a big difference in the reactivities of the α, ω-dienes was observed when $x = 0$ (vinyl type) and $x = 1$ (allyl type) (cf. Scheme 21).

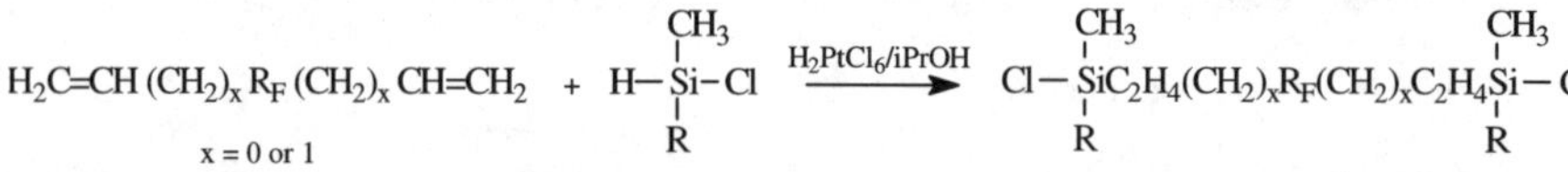

$$H_2C=CH\ (CH_2)_x\ R_F\ (CH_2)_x\ CH=CH_2\ +\ H-\underset{\underset{R}{|}}{\overset{\overset{CH_3}{|}}{Si}}-Cl\ \xrightarrow{H_2PtCl_6/iPrOH}\ Cl-\underset{\underset{R}{|}}{\overset{\overset{CH_3}{|}}{Si}}C_2H_4(CH_2)_xR_F(CH_2)_xC_2H_4\underset{\underset{R}{|}}{\overset{\overset{CH_3}{|}}{Si}}-Cl$$

x = 0 or 1

Scheme 21

The hydrosilylation, with Speier catalyst ($H_2PtCl_6/iPrOH$), was quantita-tive with allyl type α, ω-dienes, whereas with vinyl type α, ω-dienes it led to a great amount of by-products. It was thus necessary to achieve the hydrosi-lylation in the presence of a peroxide.

Hydrolysis of α, ω-bischlorosilanes issued from the hydrosilylation was quantitative, and an important amount of oligomers was already present in the compound issued from the vinyl type α, ω-diene (silicone with $x = 0$). Then, the polymerization, or polycondensation was faster when $x = 0$ and it led to a polymer of higher molecular weight.

Concerning the thermal properties of these hybrid homopolymers, the T_g was higher and the thermal stability at high temperature was lower when $x = 1$ than when $x = 0$ [48] (cf. Table 1).

Table 1 Thermal data for hybrid F/silicone homopolymers

$HO{-}[\overset{\underset{R}{|}}{\underset{|}{Si}}{-}R'{-}\overset{\underset{R}{|}}{\underset{|}{Si}}O]_n{-}H$ (CH₃, CH₃)		DSC (10 °C/min)			TGA (5 °C/min)	
		T_g	T_m	T_c	$T_{50\%}$ (N$_2$)	$T_{50\%}$ (Air)
R = CH₃	R′ = C₂H₄C₆F₁₂C₂H₄	– 53	26	– 11	470	380
	R′ = C₃H₆C₆F₁₂C₃H₆	– 40	25	– 27	465	330
R = C₂H₄CF₃	R′ = C₂H₄C₆F₁₂C₂H₄	– 28			490	410
	R′ = C₃H₆C₆F₁₂C₃H₆	– 18			465	360
R = C₂H₄C₄F₉	R′ = C₂H₄C₆F₁₂C₂H₄	– 42			490	360
	R′ = C₃H₆C₆F₁₂C₃H₆	– 29			470	310
R = CH₃	R′ = C₃H₆/HFP/C₄F₈/ HFP/C₃H₆	– 49			425	300
R = C₂H₄CF₃	R′ = C₃H₆/HFP/C₄F₈/ HFP/C₃H₆	– 34			445	310
R = C₂H₄C₄F₉	R′ = C₃H₆/C₂F₄/VDF/ HFP/C₃H₆	– 47			420	315

HFP = $-$ CF(CF₃) $-$ CF₂ $-$
VDF = $-$ CH₂ $-$ CF₂ $-$

Copolymers were obtained by copolycondensation of hybrid bis-silanols and dichlorosilanes to give random copolymers or by copolycondensation of hybrid bis-silanols and diacetamidosilanes to give alternated copolymers. The thermal properties of these two kinds of copolymers were not much different and were slightly better than those of the parent hybrid homopolymers [50].

Some of these polymeric hybrid fluorosilicones were crosslinked to obtain fluorosiloxane elastomers that combine a good flexibility at low temperature, lower than $-$ 40 °C, and a good thermooxidative stability over 250 °C [51, 52]. They may be proposed as alternative materials with respect to polyfluoroolefin elastomers.

In 1995–1996, several Japanese patents [53–56] were issued about new fluorinated silalkylene-siloxanes which were shown to exhibit a high resistance to chain-scission by acid or alkali, but nothing was said about their thermal or mechanical properties. Only their surface properties, due to fluorinated side chains, were studied.

So, we were interested in reproducing the synthesis of one of these products [57] to compare its thermal properties to those of the hybrid fluorosilicones that we had previously described. The synthesis was performed according to the following scheme:

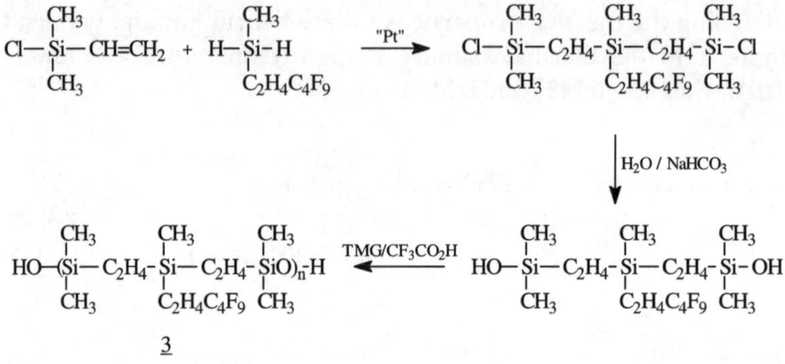

$$\underset{\overset{|}{CH_3}}{\overset{\overset{|}{CH_3}}{Cl-Si-CH{=}CH_2}} \;+\; \underset{\overset{|}{C_2H_4C_4F_9}}{\overset{\overset{|}{CH_3}}{H-Si-H}} \;\xrightarrow{\text{"Pt"}}\; \underset{\overset{|}{CH_3}}{\overset{\overset{|}{CH_3}}{Cl-Si-C_2H_4-}}\underset{\overset{|}{C_2H_4C_4F_9}}{\overset{\overset{|}{CH_3}}{Si-C_2H_4-}}\underset{\overset{|}{CH_3}}{\overset{\overset{|}{CH_3}}{Si-Cl}}$$

(↓ H_2O / $NaHCO_3$)

$$\underset{\overset{|}{CH_3}}{\overset{\overset{|}{CH_3}}{HO{-}(Si-C_2H_4-}}\underset{\overset{|}{C_2H_4C_4F_9}}{\overset{\overset{|}{CH_3}}{Si-C_2H_4-}}\underset{\overset{|}{CH_3}}{\overset{\overset{|}{CH_3}}{SiO)_n H}} \;\xleftarrow{\text{TMG/CF}_3\text{CO}_2\text{H}}\; \underset{\overset{|}{CH_3}}{\overset{\overset{|}{CH_3}}{HO-Si-C_2H_4-}}\underset{\overset{|}{C_2H_4C_4F_9}}{\overset{\overset{|}{CH_3}}{Si-C_2H_4-}}\underset{\overset{|}{CH_3}}{\overset{\overset{|}{CH_3}}{Si-OH}}$$

<u>3</u>

Scheme 22

This new fluorinated polysilalkylene-siloxane <u>3</u> presented a rather low $T_g = -65\,^\circ C$ and its thermal stability at high temperature was comparable to that of the classical polytrifluoropropylmethylsiloxane (PTFPMS), i.e. it was less stable than our previous hybrid silicones.

Finally, various Japanese patents [58–60] should be cited as they describe the synthesis of homopolymers and copolymers with a nonfluorinated backbone, issued from the corresponding bis silanol monomers and having the following formulas:

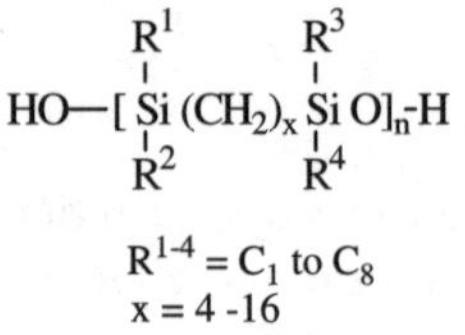

$$HO{-}[\,\underset{\overset{|}{R^2}}{\overset{\overset{|}{R^1}}{Si}}\,(CH_2)_x\,\underset{\overset{|}{R^4}}{\overset{\overset{|}{R^3}}{Si}}\,O]_n H$$

$$R^{1\text{-}4} = C_1 \text{ to } C_8$$
$$x = 4\text{ -}16$$

Fig. 15

$$XO{-}[\,\underset{\overset{|}{R^2}}{\overset{\overset{|}{R^1}}{Si}}\,(CH_2)_x\,\underset{\overset{|}{R^4}}{\overset{\overset{|}{R^3}}{Si}}\,O]_n\,(\,\underset{\overset{|}{R^6}}{\overset{\overset{|}{R^5}}{Si}}\,O\,)_m X$$

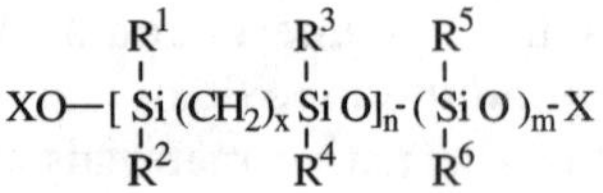

Fig. 16

with $R^{1\text{-}5}$ = monovalent substituted (or not) aliphatic hydrocarbon;

R^6 = unsaturated monovalent hydrocarbon;

$X = H$ or $SiR^7R^8R^9$ and $R^{7\text{-}9}$ = monovalent substituted (or not) hydrocarbon.

These products have been used in silicone compositions that have been crosslinked and the elastomers obtained showed very good mechanical properties (high tension and tear strength).

3
Synthesis of Hybrid Silicones Through Hydrosilylation of α, ω-Dienes. Hydrosilylation Polymerization

The principle of this method is the addition of α, ω-dienes onto α, ω-dihydrosiloxanes or oligosiloxanes according to Scheme 3 (previously given in the introduction).

The first works performed by this method were published by Russian researchers [61–63] who had studied the reaction described in Scheme 23:

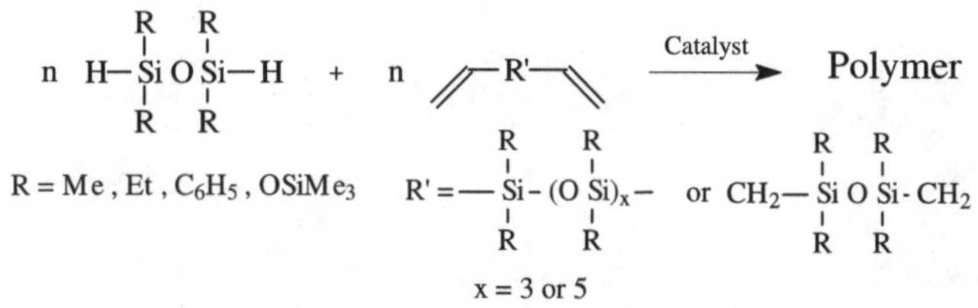

$$n \ \text{H–}\overset{\overset{R}{|}}{\underset{\underset{R}{|}}{\text{Si}}}\text{O}\overset{\overset{R}{|}}{\underset{\underset{R}{|}}{\text{Si}}}\text{–H} \ + \ n \ \diagup\!\!\!\diagdown\text{–R'–}\diagdown\!\!\!\diagup \ \xrightarrow{\text{Catalyst}} \ \text{Polymer}$$

$$R = \text{Me}, \text{Et}, \text{C}_6\text{H}_5, \text{OSiMe}_3 \quad R' = -\overset{\overset{R}{|}}{\underset{\underset{R}{|}}{\text{Si}}}\text{–}(\text{O} \ \overset{\overset{R}{|}}{\underset{\underset{R}{|}}{\text{Si}}})_x- \quad \text{or} \quad \text{CH}_2-\overset{\overset{R}{|}}{\underset{\underset{R}{|}}{\text{Si}}}\text{O} \ \overset{\overset{R}{|}}{\underset{\underset{R}{|}}{\text{Si}}}\text{-CH}_2$$

$$x = 3 \text{ or } 5$$

Scheme 23

The authors had used a Speier catalyst, $\text{H}_2\text{PtCl}_6/i\text{PrOH}$ and obtained products with low molecular weights (1000–2000).

More recently, Dvornic et al. [64, 65] used the hydrosilylation polymerization method between 1,1,3,3-tetramethyl disiloxane and 1,3-divinyl 1,1,3,3-tetramethyl disiloxane and succeeded in obtaining the first hybrid silicones, called here "polycarbosiloxanes," with a high molecular weight (up to 76 000), according to the following reaction:

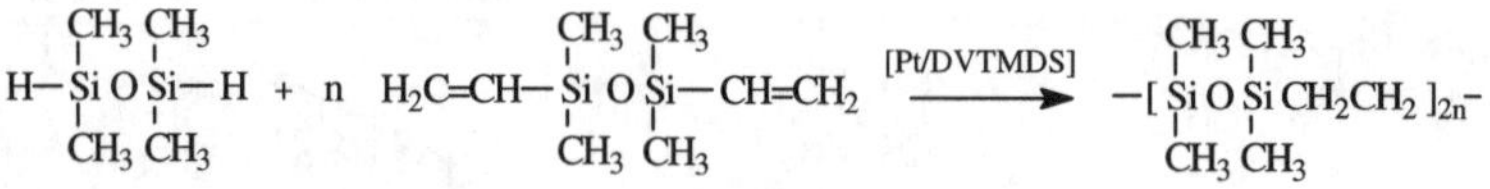

$$\text{H–}\overset{\overset{\text{CH}_3}{|}}{\underset{\underset{\text{CH}_3}{|}}{\text{Si}}}\text{O}\overset{\overset{\text{CH}_3}{|}}{\underset{\underset{\text{CH}_3}{|}}{\text{Si}}}\text{–H} \ + \ n \ \text{H}_2\text{C=CH–}\overset{\overset{\text{CH}_3}{|}}{\underset{\underset{\text{CH}_3}{|}}{\text{Si}}}\text{O}\overset{\overset{\text{CH}_3}{|}}{\underset{\underset{\text{CH}_3}{|}}{\text{Si}}}\text{–CH=CH}_2 \ \xrightarrow{\text{[Pt/DVTMDS]}} \ -[\overset{\overset{\text{CH}_3}{|}}{\underset{\underset{\text{CH}_3}{|}}{\text{Si}}}\text{O} \ \overset{\overset{\text{CH}_3}{|}}{\underset{\underset{\text{CH}_3}{|}}{\text{Si}}}\text{CH}_2\text{CH}_2 \ ]_{2n}-$$

Scheme 24

The hydrosilylation was, then, catalyzed by the complex Platinum-divinyl-1,3 tetramethyldisiloxane [Pt-DVTMDS] or Karstedt catalyst. It was studied in different conditions: in bulk, with a diluted and with a concentrated toluene solution. The higher molecular weight was obtained when the polymerization was achieved without any solvent. Actually, according to Dvornic, "the selection of Karstedt catalyst seems to be the key factor for the obtention of high molecular weights. In contrast to hexachloroplatinic acid utilized by the previous Russian workers, and that may generate HCl after reduction, the use of [Pt-DVTMDS] complex enables the hydrosilylation polymerization reaction to proceed unobstructed and to yield high molecular weight polymers."

Rheological studies and thermogravimetric analysis of the obtained polymers showed that the flexibility, the thermal and oxidative stabilities were lower than for polysiloxanes with a close structure. This is due to the stiffening and destabilizing effect of the C – C groups introduced between the main Si – O – Si units of the chain.

However, these authors strongly insisted on the fact that hydrosilylation is a good method for the preparation of linear carbosiloxanes with high molecular weights.

Very recently, another example of [Pt-DVTMDS] catalyzed hydrosilylation copolymerization leading to fluorinated copoly(carbosiloxane)s has been described [66]. It consisted of the addition of α, ω-divinyl fluorooligosiloxanes onto α, ω-dihydro fluorooligosiloxanes as follows:

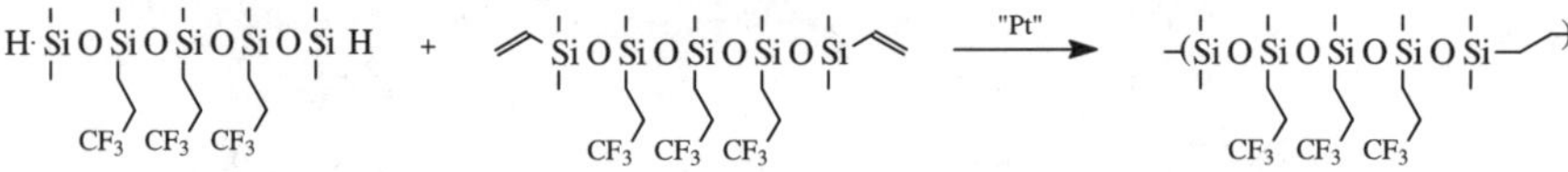

Scheme 25

The structures of the copoly(carbosiloxane)s have been determined by I.R. as well as by ^{1}H, ^{13}C, ^{19}F and ^{29}Si NMR spectroscopy. The GPC analysis showed that high molecular weights were obtained (20 000–40 000) and the DSC and TGA analyses showed very low T_gs, in the range – 77 to – 80 °C and a good thermal stability both in nitrogen (stability to approximately 380 °C) and in air (stability to approximately 270 °C).

Another example of polyhydrosilylation is the addition of diallyl bisphenol A to tetramethyldisiloxane which was reported by Lewis and Mathias in 1993 [67, 68] (cf. Scheme 26):

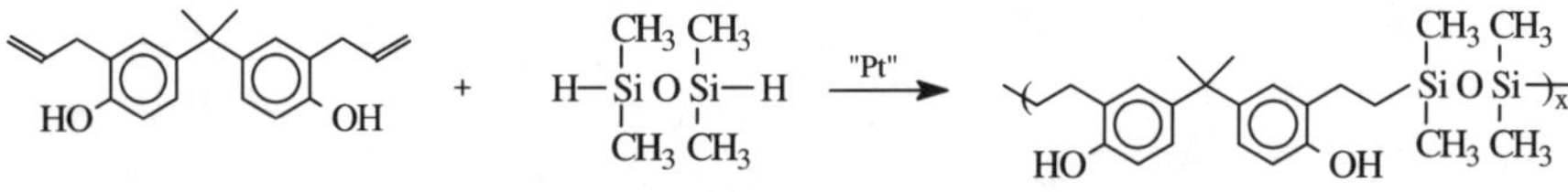

Scheme 26

The reaction is strongly exothermic and must be performed in a solvent as the co-reagents are not miscible. But, even if the reaction is performed at 0 °C, the molecular weights are here limited by the nonstoichiometry due to the volatility of the disiloxane.

Some years later, almost the same reaction was performed with a hexafluoro derivative of bisphenol A [69, 70] and the resulting polymers proved to be excellent sorbents for basic vapors due to their strong hydrogen bond acidity.

Recently, Boileau et al. [71, 72] performed the polyhydrosilylation of diallyl bisphenol A with hydride terminated polydimethylsiloxanes to prepare "tailor-made polysiloxanes with anchoring groups" composed of dimethylsiloxane segments (DMS) of different lengths, regularly separated by one bisphenol A (BPA) unit. They studied the influence of the control of the

[Si – H]/[double bond] ratio and the protection of the – OH groups on the molecular weight distribution of the polymers. A strong influence of the DMS segment length and of the presence of H-bonding interactions on the thermal properties of the resulting polymers was observed. The T_g decreased (from + 32 to – 114 °C) when increasing the siloxane segment length and the TGA analysis under nitrogen showed a quite good thermal stability.

The polyhydrosilylation method had also been applied earlier by Boileau et al. [73] to synthesize well-defined polymers containing silylethylene siloxy units (cf. Figs. 17, 18 and 19):

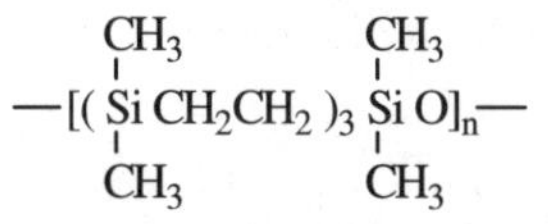

$$-[(\underset{\underset{CH_3}{|}}{\overset{\overset{CH_3}{|}}{Si}}\,CH_2CH_2\,)_3\,\underset{\underset{CH_3}{|}}{\overset{\overset{CH_3}{|}}{Si}}\,O]_n-$$

Fig. 17

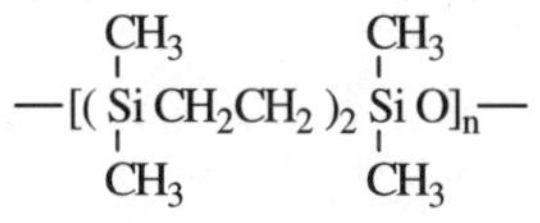

$$-[(\underset{\underset{CH_3}{|}}{\overset{\overset{CH_3}{|}}{Si}}\,CH_2CH_2\,)_2\,\underset{\underset{CH_3}{|}}{\overset{\overset{CH_3}{|}}{Si}}\,O]_n-$$

Fig. 18

$$-[\,\underset{\underset{CH_3}{|}}{\overset{\overset{CH_3}{|}}{Si}}\,CH_2CH_2\,\underset{\underset{CH_3}{|}}{\overset{\overset{CH_3}{|}}{Si}}\,O]_n-$$

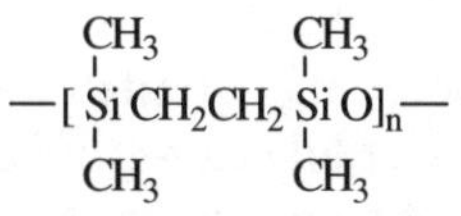

Fig. 19

Additionally, the method has been used in a patent to prepare poly-(imidesiloxanes) (PI/PS) "in a relatively simple manner, without undesirable side reactions and in which high conversions are achieved in short reaction times" [74]. They reacted an N,N'-dialkenyldiimide with an organosilicon compound containing two Si – H, in the presence of diCpPtCl$_2$ as catalyst (cf. Scheme 27).

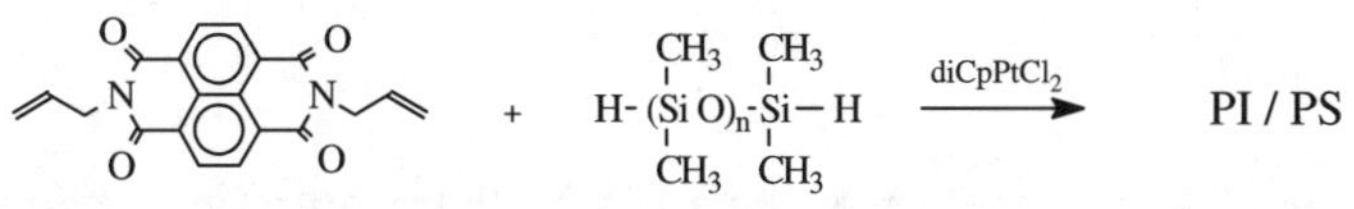

Scheme 27

The prepared poly-(imidesiloxanes) showed higher heat stability and their T_g was lower when the proportion of siloxane was higher. These products may find applications as coatings, as adhesives or as membranes for gas separation.

The same method was used to prepare thermoplastic siloxane elastomers based on poly(arylenevinylenesiloxanes) compounds [75]. The polyhydrosilylation was then performed between an α, ω-dialkenylarylenevinylene and an organosilicon compound containing two Si–H, in the presence of diCpPtCl$_2$ as shown in Scheme 27.

More recently, we have also reported the synthesis of thermoplastic siloxane elastomers based on hybrid polysiloxane/polyimide block copolymers (the hybrid polysiloxane being fluorinated or not) that were obtained through polyhydrosilylation of dienes with α, ω-dihydrooligosiloxanes [76–78], as follows:

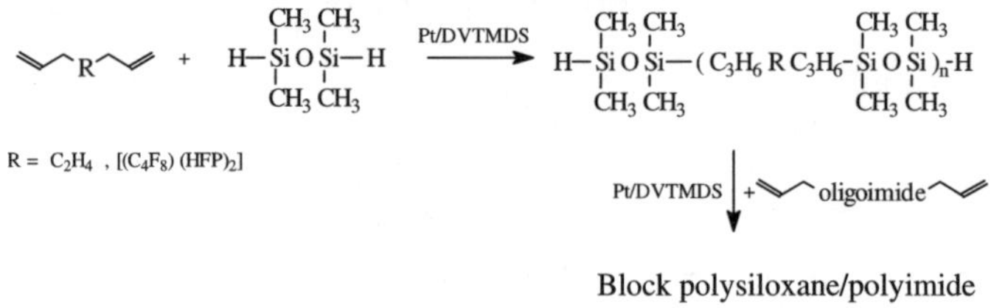

$R = C_2H_4$, $[(C_4F_8) (HFP)_2]$

Block polysiloxane/polyimide

Scheme 28

These block copolymers exhibited both good thermomechanical properties and low surface tension and some of them exhibited also thermoplastic elastomers properties.

As a variant to this method, it may be cited the obtention of block copolymers through hydrosilylation of allyloxy-4 benzaldehyde with α, ω-dihydro oligosiloxanes in the presence of a Pt catalyst [79] (cf. Scheme 29):

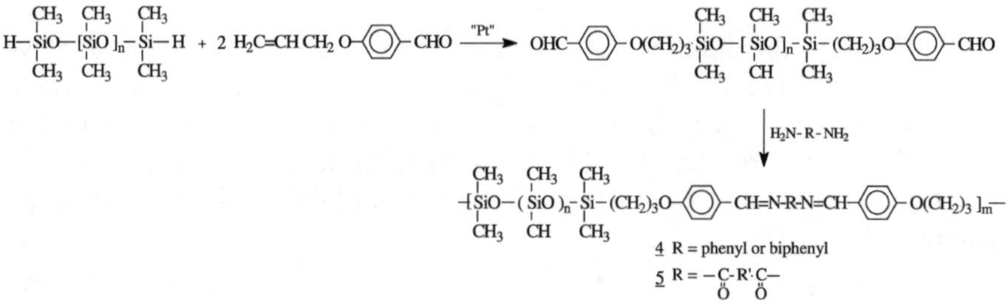

$\underline{4}$ R = phenyl or biphenyl
$\underline{5}$ R = –C-R'-C–
            Ö Ö

Scheme 29

These block copolymers may be used as thermoplastic materials or as additives, in the case of compounds $\underline{5}$, as they may be incorporated into a polyamide matrix.

The polyhydrosilylation method has also been used in an American patent [80] and a Japanese patent [81] to obtain hybrid silicone copolymers.

The former describes the hydrosilylation of trienes (only on the terminal unsaturated groups) by hydrosiloxanes, to give polysilalkylene siloxanes (cf. Scheme 30):

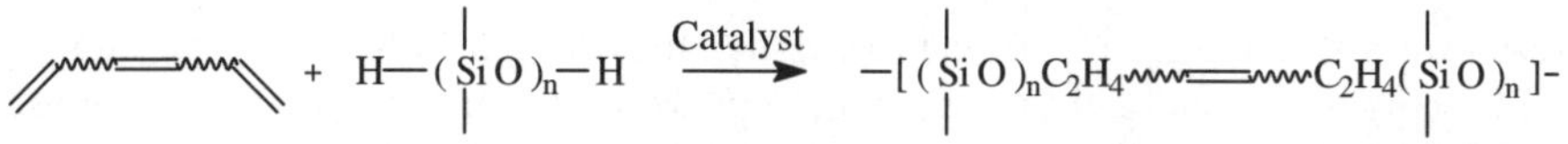

Scheme 30

The latter describes vulcanized silicone rubbers exhibiting very good mechanical resistances and obtained starting from hybrid silicone copolymers prepared via hydrosilylation of dimethyl silyl vinyl ended siloxanes with poly dimethyl methyl hydrogeno siloxanes, in the presence of a Pt catalyst (cf. Scheme 31):

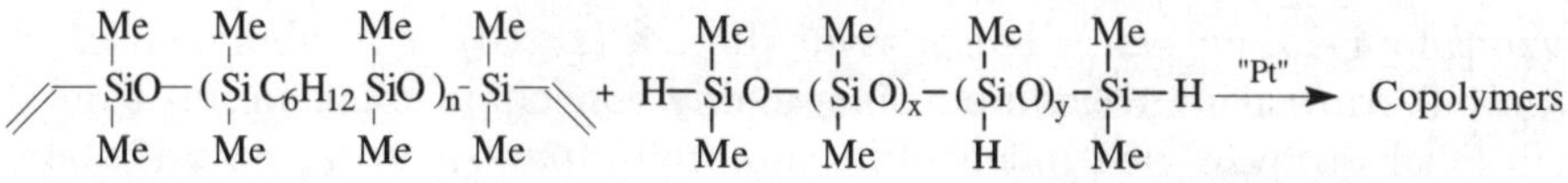

Scheme 31

Finally, the platinum-catalyzed hydrosilylation polymerization was also used very recently by Cassidy et al. [82, 83] to prepare fluorine containing "silicon-organic hybrid polymers" in supercritical carbon dioxide ($ScCO_2$) (cf. Scheme 32).

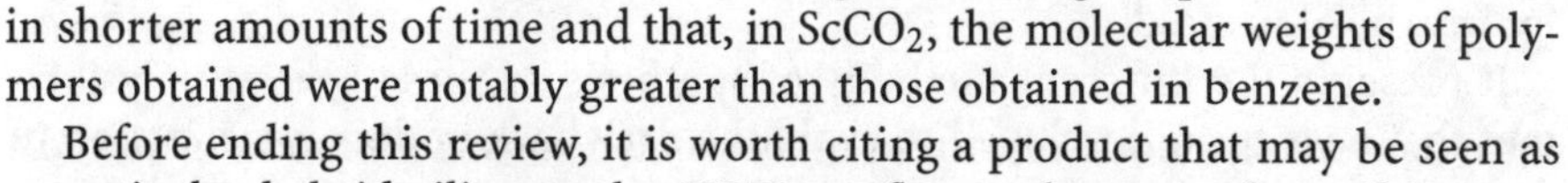

Scheme 32

They showed that the $ScCO_2$ reaction provided higher percent conversion in shorter amounts of time and that, in $ScCO_2$, the molecular weights of polymers obtained were notably greater than those obtained in benzene.

Before ending this review, it is worth citing a product that may be seen as a particular hybrid silicone: the SIFEL perfluoro elastomer from Shin-Etsu. Actually, it consists of a perfluoroether polymer backbone combined with an addition-curing silicone crosslinker. The perfluoroether polymer was capped with vinyl silicone functions and the crosslinking was achieved with a special cross-linker containing several Si – H end groups (general type as in Fig. 20), in the presence of a platinum catalyst [84, 85].

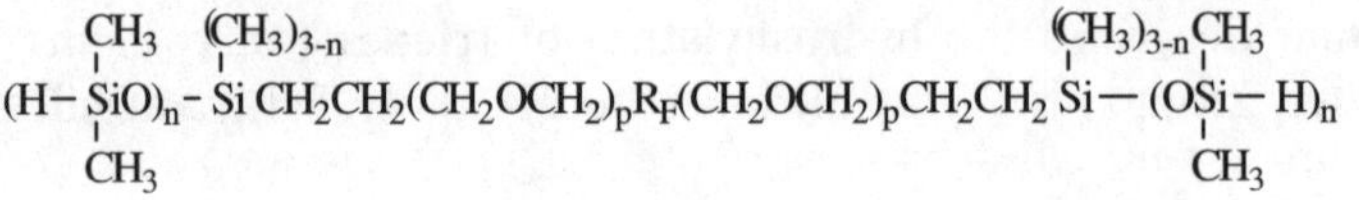

$$\underset{\overset{|}{CH_3}}{\overset{\overset{|}{CH_3} \quad (CH_3)_{3-n}}{(H-SiO)_n}} - Si\,CH_2CH_2(CH_2OCH_2)_pR_F(CH_2OCH_2)_pCH_2CH_2\,Si \underset{\overset{|}{CH_3}}{\overset{(CH_3)_{3-n}CH_3}{- (OSi-H)_n}}$$

R_F = perfluoropolyether or perfluoroalkylene group

n = 1, 2 or 3

p = 0 or 1

Fig. 20

The product is described as a liquid perfluoroelastomer and it is becoming popular in the industries as a universal material for O-rings, diaphragms and other mold parts due to its unique properties issued from its special chemical formula (cf. Fig. 21):

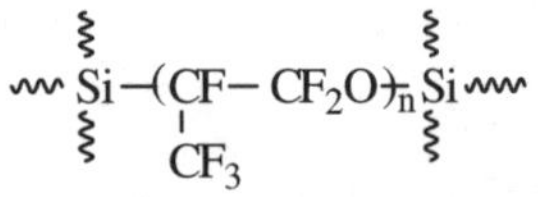

$$\text{---}\,\overset{\xi}{\underset{\xi}{Si}}\text{---}(\underset{\overset{|}{CF_3}}{CF} - CF_2O)_n\,\overset{\xi}{\underset{\xi}{Si}}\,\text{---}$$

Fig. 21

The compound is specially interesting for aerospace industries as it can perform well for different media: jet fuel, hydraulic oil, engine oil and hydraulic fluid, under severe environmental conditions.

This new type of elastomer, with its wide range of applications, constitutes a solution to some of the increasingly complex demands of the different industries.

4
Conclusions

This review on hybrid silicones does not pretend to be an exhaustive list of all the polysilalkylene or polysilarylene siloxanes, fluorinated or not, that have been reported in the literature and that may also be called "polycarbosiloxanes."

It presents the different methods of synthesis of these special polysiloxanes that have been developed to avoid the drawback of depolymerization of classical polysiloxanes in certain conditions of temperature or of acid or base medium.

The first method that has been mainly used since the 1960s was based on polycondensation of α, ω-dihydroxysiloxanes, while the second method which has been developing during the last three decades is based on polyhydrosilylation of α, ω-diolefines with α, ω-dihydro terminated siloxanes.

All the homopolymers or copolymers that have been obtained show very interesting properties in terms of thermal stability. They generally present rather low T_gs and good stability at high temperature and may thus be used over a wide range of temperature. Furthermore, in the search for new materials for new applications, the obtention of polymers with specific properties is required, and depending on the nature of their main chain (alkyl, fluoroalkyl, aryl, fluoroaryl, alkyl ether, etc.) and on the nature of their side chains, these hybrid silicones may be directed to exhibit specific properties.

Actually, a few years ago, Hergenrother [86] stated the precise requirements of the technology for high speed civil transports (HSCTs): the sealants must exhibit a combination of properties such as elongation, moderate peel strength, fuel resistance and performance for 60 000 h at 177 °C. He said that the most popular commercially available fuel tank sealant that can be used at a temperature of around 177 °C is based upon poly(3,3,3-trifluoropropyl methylsiloxane), but this product may degrade after continued exposure to high temperature.

Since then, the Sifel from Shin-Etsu has emerged, but it is a very expensive material.

So, finding a good combination of hybrid or silalkylene siloxanes, classical siloxanes, silarylene siloxanes, preferably fluorinated, remains a challenge to obtain the best elastomer.

It seems that there is still a promising future for these hybrid silicone materials.

References

1. Thomas TH, Kendrick TC (1969) J Polym Sci A7:537
2. Grassie N, MacFarlane IG (1978) Eur Polym J 14:875
3. Sommer LH, Ansul G (1955) J Am Chem Soc 77:2482
4. Merker RL, Scott MJ (1964) J Polym Sci A2:15
5. Merker RL, Scott MJ, Haberland GG (1964) J Polym Sci A 2:31
6. Curry JE, Byrd JD (1965) J Appl Polym Sci 9:295
7. Burks RE Jr, Covington ER, Jackson MV, Curry JE (1973) J Polym Sci Polym Chem 11:319
8. Dvornic PR, Lenz RW (1980) Polym Prepr J Am Chem Soc 21:142
9. Dvornic PR, Lenz RW (1980) J Appl Polym Sci 25:641
10. Dvornic PR, Lenz RW (1982) J Polym Sci Polym Chem 20:593 and 951
11. Lai YC, Dvornic PR, Lenz RW (1982) J Polym Sci Polym Chem 20:2277
12. Livingston ME, Dvornic PR, Lenz RW (1982) J Appl Polym Sci 27:3239
13. Dvornic PR, Lenz RW (1983) Polymer 24:763
14. Dvornic PR, Perpall HJ, Uden PC, Lenz RW (1989) J Polym Sci Polym Chem 27:3503
15. Dvornic PR (1992) Polym Bull 25:339
16. Dvornic PR, Lenz RW (1992) Macromolecules 25:3769
17. Dvornic PR, Lenz RW (1994) Macromolecules 27:5833
18. Hani R, Lenz RW (1990) In: Ziegler JM, Fearon FWG (eds) Silicon-based Polymer

Science Advances in Chemistry Series 224. Am Chem Soc, Washington, D.C., p 741
19. Zhu HD, Kantor SW, McKnight WJ (1998) Macromolecules 31:850
20. Lauter U, Kantor SW, McKnight WJ (1998) Polym Prep J Am Chem Soc 39:613
21. Lauter U, Kantor SW, Schmidt-Rohr K, McKnight WJ (1999) Macromolecules 32:3426
22. Williams EA, Wengrovius JH, VanValkenburgh VM, Smith JF (1991) Macromolecules 24:1145
23. Boutevin B, Pietrasanta Y, Youssef B (1988) J Fluorine Chem 39:61
24. Boutevin B, Youssef B (1989) J Fluorine Chem 45:61
25. Benouargha A, Boutevin B, Caporiccio G, Essassi E, Guida-Pietrasanta F, Ratsimihety A (1997) Eur Polym J 33:1117
26. Stern SA, Shah VM, Hardy BJ (1987) J Polym Sci Polym Phys 25:1263
27. Zhang R, Pinhas AR, Mark JE (1997) Polym Prep J Am Chem Soc 38:298
28. Homrighausen CL, Keller TM (2000) Polym Mater Sci Eng 83:8
29. Buvat P, Jousse F, Nony F, Gerard JF (2003) WO 2003076516 (CEA, France); CA (2003) 139:246474
30. Nony F (2003) 14 November 2003, PhD Thesis, INSA Lyon
31. Kitamura N, Yamamoto T (2003) Appl Organomet Chem 17:840
32. Loree LA, Brown ED (1970) FR 2035607 (Dow Corning Corp., USA); CA (1971) 74:88528
33. Grindahl GA (1970) DE 2007940 (Dow Corning Corp., USA); CA (1970) 73:121388
34. Critchley JP, MacLoughlin VCR, Thrower J, White IM (1970) Br Polym J 2:288
35. Cotter JL, Knight GJ, Wright WW (1975) Br Polym J 7:381
36. Patterson WJ, Morris DE (1972) J Polym Sci Polym Chem 10:169
37. Pittman CU Jr, Patterson WJ, McManus SP (1976) J Polym Sci Polym Chem 14:1715
38. Loughran GA, Griffin WR (1985) Polym Prep J Am Chem Soc 26:150
39. Rosenberg H, Eui-Won Choe (1979) Organ Coat Plastic Preprints 40:792
40. Rizzo J, Harris FW (2000) Polymer 41:5125
41. Kim YK (1971) Rubber Chem Technol 1350
42. Kim YK, Pierce OR, Bourrie DE (1972) J Polym Sci Polym Chem 10:947
43. Pierce OR , Kim YK (1973) Appl Polym Symp 22:103
44. Kim YK, Riley MO (1976) US Patent 3975362 (Dow Corning Corp USA)
45. Riley MO, Kim YK, Pierce OR (1977) J Fluorine Chem 10:85
46. Riley MO, Kim YK, Pierce OR (1978) J Polym Sci Polym Chem 16:1929
47. Boutevin B, Guida-Pietrasanta F, Ratsimihety A, Caporiccio G (1996) US Patent 5527933; CA (1996) 125:115552
48. Ameduri B, Boutevin B, Guida-Pietrasanta F, Manseri A, Ratsimihety A, Caporiccio G (1996) J Polym Sci Polym Chem 34:3077
49. Boutevin B, Caporiccio G, Guida-Pietrasanta F, Ratsimihety A (1997) Recent Res Devel Polymer Sci 1:241
50. Boutevin B, Caporiccio G, Guida-Pietrasanta F, Ratsimihety A (1998) Macromol Chem Phys 199:61
51. Boutevin B, Caporiccio G, Guida-Pietrasanta F, Ratsimihety A (2001) EP 1097958 (Daikin Ind); CA (2001) 134:341498
52. Boutevin B, Caporiccio G, Guida-Pietrasanta F, Ratsimihety A (2003) J Fluorine Chem 124:131
53. Kobayashi H (1995) EP 665270 (Dow Corning Toray Silicone Japan); CA (1995) 123:288221
54. Kobayashi H (1996) EP 690088 (Dow Corning Toray Silicone Japan); CA (1996) 124:177296
55. Hamada Y, Kobayashi H, Nishiumi W (1996) EP 699725 (Dow Corning Toray Silicone

Japan); CA (1996) 124:292544
56. Kobayashi H (1996) EP 702048 (Dow Corning Toray Silicone Japan); CA (1996) 124:318198
57. Boutevin B, Guida-Pietrasanta F, Ratsimihety A, Caporiccio G (1997) Main Group Metal Chem 20:133
58. Takaai T, Kinami H, Sato S, Matsuda T (1993) JP 05171048 A2 (Shin Etsu Chem Ind Co Japan); CA (1994) 120:109289
59. Kishita H, Sato S, Yamaguchi K, Koike N, Matsuda T (1993) EP 556780 A2 (Shin Etsu Chem Ind Co Japan); CA (1994) 120:193828
60. Takago T, Sato S, Koike N, Matsuda T (1993) EP 549214 (Shin Etsu Chem Ind Co Japan); CA (1993) 119:227830
61. Petrov AD, Vdovin VM (1959) Izv Akad Nauk SSSR 939
62. Andrianov KA, Kocetkova AC, Hananashvili LM (1968) Zh Obsch Khim 38:175
63. Andrianov KA, Gavrikova LA, Rodionova EF (1971) Vysokomol Soedin A13:937
64. Dvornic PR, Gerov VV, Govedarica MN (1994) Macromolecules 27:7575
65. Dvornic PR, Gerov VV (1994) Macromolecules 27:1068
66. Grunlan MA, Mabry JM, Weber WP (2003) Polymer 44:981
67. Lewis CM, Mathias LJ (1993) Polym Prep J Am Chem Soc 34:491
68. Mathias LJ, Lewis CM (1993) Macromolecules 26:4070
69. Grate JW, Kaganove SN, Patrash SJ, Craig R, Bliss M (1997) Chem Mater 9:201
70. Kaganove SN, Grate JW (1998) Polym Prep J Am Chem Soc 39:556
71. Tronc F, Lestel L, Boileau S (1998) Polym Prep J Am Chem Soc 39:583
72. Tronc F, Lestel L, Boileau S (2000) Polymer 41:5039
73. Jallouli A, Lestel L, Tronc F, Boileau S (1997) Macromol Symp 122:223
74. Wenski G, Maier L, Kreuzer FH (1990) US Patent 5009934 (Consortium für Elektrochemische Ind GmbH); CA (1991) 114:248026
75. Funk E, Kreuzer FH, Gramshammer C, Lottner W (1993) US5185419 (Consortium für Elektrochemische Ind GmbH); CA (1991) 115:280814
76. Andre S, Guida-Pietrasanta F, Ratsimihety A, Rousseau A, Boutevin B (2000) Macromol Chem Phys 201:2309
77. Andre S, Guida-Pietrasanta F, Rousseau A, Boutevin B (2001) J Polym Sci Polym Chem 39:2414
78. Andre S, Guida-Pietrasanta F, Rousseau A, Boutevin B, Caporiccio G (2002) J Polym Sci Polym Chem 40:4485
79. Madec PJ, Marechal EJ (1993) Polym Prep J Am Chem Soc 34:814
80. Durfee LD, Hilty TK (1993) EP 539065 (Dow Corning Corp USA); CA (1993) 119:272012
81. Takago T, Sato S, Noike N, Matsuda T (1993) EP 549214 A2 (Shin Etsu Chem Ind Co Japan); CA (1993) 119:227830
82. Green JW, Rubal MJ, Osman BM, Welsch RL, Cassidy PE, Fitch JW, Blanda MT (2000) Polym Adv Technol 11:820
83. Hui Zhou, Venumbaka SR, Fitch JW III, Cassidy PE (2003) Macromol Symp 192:115
84. Uritani P, Kishita H (2002) High Performance Elastomers 2002, 13–14 November 2002, Cologne, Germany
85. Waksman L, Kishita H, Sato S, Tarumi Y (2001) Society of Automotive Engineers, Special Publication, SP-1611:47
86. Hergenrother PM (1996) Trend Polym Sci 4:104

Editor: Oskar Nuyken

Adv Polym Sci (2005) 179: 29–82
DOI 10.1007/b107204
© Springer-Verlag Berlin Heidelberg 2005
Published online: 6 June 2005

Epoxy Layered Silicate Nanocomposites

Ole Becker · George P. Simon (✉)

Department of Materials Engineering, Monash University,
Clayton, 3800 Victoria, Australia
ole.becker@airbus.com, george.simon@eng.monash.edu.au

1	Introduction	31
2	Current Modifications of Epoxies	33
2.1	Particulate Toughening of Thermosets	33
2.2	Rubber Toughening of Thermosets	33
2.3	Thermoplastic toughening of thermosets	34
2.4	Epoxy Fibre Composites	35
3	Crystallography and Surface Modification of Layered Silicates	36
4	Characterization of Thermosetting Layered Silicate Nanocomposite Morphology	38
4.1	Wide-angle X-ray diffraction	39
4.2	Small angle X-ray Diffraction (SAXD)	40
4.3	Transmission electron microscopy (TEM)	41
4.4	Optical and Scanning Electron Microscopy (SEM)	42
4.5	Atomic Force Microscopy (AFM)	42
4.6	NMR Dispersion Measurements of Nanocomposites	43
5	Synthesis of Thermosetting Layered Silicate Nanocomposites	44
6	Controlling the Morphology of Epoxy Nanocomposites	45
6.1	Mechanism of clay dispersion	45
6.2	The Nature of the Silicate and the Interlayer Exchanged Ion	49
6.3	Curing agent	52
6.4	Cure Conditions	53
6.5	Other Strategies for Improved Exfoliation	54
7	Properties of Thermosetting Nanocomposites	55
7.1	Cure Properties	55
7.2	Thermal Relaxations	58
7.3	Mechanical Properties	61
7.3.1	Flexural, Tensile and Compressive Properties	61
7.3.2	Fracture Properties	63
7.4	Dimensional Stability	65
7.5	Water Uptake and Solvent Resistance	65

7.6   Thermal Stability and Flammability   66
7.7   Optical Properties   69

8   **Ternary Layered Silicate Nanocomposite Systems**   70
8.1   Epoxy fiber nanocomposites   70
8.2   Ternary systems consisting of a layered silicate,
       epoxy and a third polymeric component   71

9   **Conclusions and Future Directions**   75

References   77

Abstract Nanostructured organic-inorganic composites have been the source of much attention in both academic and industrial research in recent years. Composite materials, by
definition, result from the combination of two distinctly dissimilar materials, the overall behavior determined not only by properties of the individual components, but by the
degree of dispersion and interfacial properties. It is termed a *nano*composite when at
least one of the phases within the composite has a size-scale of order of nanometers.
Nanocomposites have shown improved performance (compared to matrices containing
more conventional, micron-sized fillers) due to their high surface area and significant
aspect ratios – the properties being achieved at much lower additive concentrations compared to conventional systems.

In this article, recent developments in the formation and properties of epoxy layered
silicate nanocomposites are reviewed. The effect of processing conditions on cure chemistry and morphology is examined, and their relationship to a broad range of material
properties elucidated. An understanding of the intercalation mechanism and subsequent
influences on nanocomposite formation is emphasized. Recent work involving the structure and properties of ternary, thermosetting nanocomposite systems which incorporate
resin, layered silicates and an additional phase (fibre, thermoplastic or rubber) are also
discussed, and future research directions in this highly active area are canvassed.

Keywords Nanocomposite · Epoxy · Montmorillonite · Clay · Layered silicate ·
Nanoparticle

Abbreviations

3D	three-dimensional
3DCM	3, 3'-dimethyl-4, 4'-diaminodicyclohexylmethane
μm	micrometers (10^{-6} m)
Å	angstroms (10^{-10} m)
AFM	atomic force microscopy
BDMA	benzyldimethylamine
BTFA	boron trifluoride monoethylamine
°C	degrees celcius
CEC	cation exchange capacity
CTBN	carboxy-terminated butadiene nitrile rubbers
DDS	4, 4'-diaminodiphenyl sulphone
DDM	4, 4'-diaminodiphenylmethane,

DETDA	diethyltoluenediamine (ETHACURE® 100)
DGEBA	diglycidyl ether of bisphenol A
DSC	differential scanning calorimetry
DMBA	N,N-dimethylbenzylamine
DMTA	dynamic mechanical thermal analysis
e-beam	electron beam
G	Gibb's free energy
G_{IC}	fracture energy
H	enthalpy
HBP	hyperbranched polymers
HHPA	hexahydrophthalic anhydride
HRR	heat release rate
IPNs	interpenetrating polymer networks
Jeffamine	poly(oxypropylene) diamines
K	Kelvin
m	metres
MPDA	1,4-diaminobenzene
MTHPA	methyltetrahydrophthalic anhydride -
NMA	nadic methyl anhydride (NMA)
NMR	nuclear magnetic resonance
nm	nanometers (10^{-9} m)
Nylon 6	caprolactam-based polyamide
PACM	4, 4′-diaminodicyclohexylmethane bisparaaminocyclohexylmethane
PMMA	poly(methyl methacrylate)
q	scattering vector
RFI	resin film infusion
s	seconds
S	entropy
SAXD	Small angle X-ray Diffraction
SEM	scanning electron microscopy
TEM	transmission electron microscopy
TGDDM	tetraglycidyl ether of 4,4′-diaminodiphenylmethane
T_g	glass transition temperature
TGA	thermogravimetric analysis
TGAP	triglycidyl p-amino phenol
θ	scattering angle
WAXD	wide-angle X-ray diffraction
WW	wet winding

1
Introduction

In recent years the incorporation of low concentrations of nanometer-sized fillers has become an important strategy to improve and diversify polymeric materials. A polymer nanocomposite can be defined as a two-phase system, where at least one dimension of the reinforcing filler is on the nanometer scale. Nanocomposites can vary from the inclusion of isodimensional fillers such as nanopowders, where all three dimensions are on a nanometer scale, to two-dimensional materials, such as nanorods, nanowires or nanotubes. With a thickness of the individual platelets of only 9.8 Å and an aspect ratio of up to 1000, layered silicate polymer composites are a form of nanocomposite where only the thickness is of the nanometer scale.

Clay minerals have been used for a long time as catalysts, adsorbents [1] and rheological modifiers [2, 3] in the chemical and coatings industries. The use of clays as polymer additives also has a significant history [4–6] with polymer intercalation of montmorillonite being first investigated more than 40 years ago using methyl methacrylate and montmorillonite [7]. However, it is only since the pioneering work by Toyota researchers with clays and polyamides [8–11] that layered silicates have gained importance as modifiers in improving polymer performance. The significant feature of layered silicates, in comparison to other, more commonly used fillers, is their high aspect ratio and their ability to be readily dispersible on a nanometer scale.

As illustrated in Fig. 1, layered silicate composite structures fall into three different classes: (a) microcomposites with no interaction between the clay galleries and the polymer, (b) intercalated nanocomposites, where the silicate is well-dispersed in a polymer matrix with polymer chains inserted into the galleries between the parallel, silicate platelets, and (c) exfoliated nanocomposites with fully separated silicate platelets individually dispersed or delaminated within the polymer matrix [12]. However, these terms describe only ideal cases and most observed morphologies fall between the extremes. A more detailed nomenclature will be presented later in this review.

As most work reported to date on thermosetting layered silicate nanocomposites involves epoxy resins, this review will focus on this class of thermosetting materials. However, some work published on other thermosets such as vinyl ester resins and unsaturated polyesters will be included where appropriate.

(a)   **(b)**   **(c)**

Fig. 1a–c Schematic illustration of different possible structures of layered silicate polymer composite: (**a**) microcomposite (**b**) intercalated nanocomposite (**c**) exfoliated nanocomposite [12]

2
Current Modifications of Epoxies

Epoxy thermosets are used in a variety of applications, such as coatings, adhesives, electronics or in composites in the transportation industry. Although the polyfunctional reactivity of most epoxy systems leads to a high crosslink density and the required matrix rigidity, brittleness of these materials can be problematic. In most applications the polymer is thus combined with at least one other phase, such as short or long fibres (carbon, graphite, glass or Kevlar) or a rubbery phase for toughening. The commonly-used additives for toughening of thermosets are briefly reviewed below.

2.1
Particulate Toughening of Thermosets

Rigid fillers of micron dimension, be they inorganic particles or glass beads, have long been used to reinforce thermoset materials and their behaviour is well-known [13]. They are clearly effective in terms of modulus-increase, but have also been found to lead to a concomitant improvement in fracture toughness. For example, it has been reported in an epoxy system that the addition of 40 vol % of glass beads of size between 4 and 60 μm was found to cause a two-fold increase in modulus, and a four-fold increase in critical stress intensity factor (a measure of resistance to crack growth) [14]. A number of these properties may be further enhanced by appropriate surface treatments of the particles, but this is not always the case. In terms of crack growth, toughening mechanisms are generally thought to range from encouraging of plastic deformation via stress concentration, to crack pinning which causes bowing of the crack front. The degree to which these various mechanisms influence crack propagation also depends on factors such as testing rate and temperature.

2.2
Rubber Toughening of Thermosets

Elastomeric modification is the most common way to toughen thermosetting systems. Of all the categories of rubbers studied including reactive butadiene-acrylonitrile rubbers, polysiloxanes, fluoroelastomers and acrylate elastomers, it is carboxy-terminated butadiene nitrile rubbers (CTBNs) that have shown the greatest benefits [15] and are the most widely used. The major disadvantage in rubber-toughened thermosets is that some of the beneficial properties of the thermoset matrix such as high glass transition temperature, yield strength and modulus are compromised through the incorporation of

Table 1 Change of mechanical properties of a rubber-toughened epoxy system as a function of rubber concentration [16]

Rubber [%]	Tensile strength [MPa]	Tensile Modulus [GPa]	Toughness [kJ]
0	6.92	276.1	1.40
3	6.08	318.0	4.57
6	5.19	290.0	3.31
9	4.75	234.7	2.94
12	4.10	210.3	3.00

Becker, Simon

rubber. Table 1 illustrates the dependence of tensile strength and modulus on rubber addition in a toughened epoxy resin system [16].

In rubber toughening, it is particularly important that the rubber and resin blend develops a two-phase morphology during the crosslinking reaction, where the precipitated rubber particles become dispersed in (and preferably bonded to) the resin matrix. The amount of rubber required is usually limited to concentrations of 10–15% to ensure that the rubber remains as the dispersed phase. Higher rubber concentrations would lead to phase inversion, resulting in a significant decrease in strength and stiffness. For the same reason, the cure profile must be adjusted to optimize the overall morphology, and resulting material performance. Any soluble rubber remaining in the matrix plasticises the polymer network, decreasing the glass transition temperature and modulus.

A more recent strategy to toughen thermosetting systems is through the incorporation of hyperbranched polymers (HBP), particularly those that are epoxy-terminated. Hyperbranched or dendritic type polymers are a new class of three dimensional, synthetic molecule produced by a hybrid synthetic process that generates highly branched, polydisperse molecules with novel molecular architecture. The use of HBP has shown some promising improvement in mechanical properties of epoxy systems, along with beneficial low viscosities for ease of processing [17, 18].

2.3
Thermoplastic toughening of thermosets

Although the first attempts of thermoset toughening through thermoplastic addition showed only modest enhancement in toughness [19], these studies created much interest in the field, resulting in the exploration of many different factors which lead to further significant improvements. The main areas explored were the toughening effect of reactive end-groups, morphology and matrix ductility, as well as the chemical structure and molecular weight of the thermoplastic. In brief, the key factors were found to be [20]:

Reactive endgroups	although there is incomplete agreement in the literature, the use of reactively-terminated endgroups appears desirable.
Morphology	phase-inverted or co-continuous morphologies lead to optimum toughness (not the case in rubber-toughened systems).
Matrix ductility	thermoplastic additives have been found to toughen highly crosslinked resin/amine systems more effectively than low crosslink density resins, again not found in rubber-toughened epoxy resins.
Thermoplastic structure	polymers with good thermal stability are required. The thermoplastic should be soluble in the unreacted resin but must phase separate well during cure, so as to form a clear, binary system.
Molecular weight	the toughness of the blend increases with increasing thermoplastic molecular weight due to the improved mechanical properties of the thermoplastic phase dominating blend properties

2.4
Epoxy Fibre Composites

The production of composites from epoxy resins and fibres has significantly increased in recent time. Both the fiber and polymeric phases retain their original chemical and physical identities, with mechanical properties sometimes exceeding those of the constituents. The nature of the interface of the two phases is of enormous importance, particularly where high resistance to failure is sought [21].

In high performance composites, the fibre phase is usually carbon, graphite or glass and may be short, long and aligned or woven. Intercorporation of these fibres into the epoxy matrix yields high modulus and strength, although possibly low ductility. This can lead to problems in terms of reduced impact strength at low velocities and low delamination resistance with out-of-plane strength being poor [21]. Problematically, such damage can be sub-surface and remain undetected, reducing material performance. Improving the intrinsic *matrix* toughness can alleviate this to some degree but such strategies are not as effective in toughening composites. Two-dimensional structures usually offer good properties in the laminate plane, with more recent research focusing on laminate improvements via more three-dimensional (3D) structures [22, 23]. Such 3D laminates are found to encourage fibre debonding and micro-cracking, as well as resisting crack growth between layers. 3D composites can involve processes such as weaving, knitting and stitching but this requires special fabrication techniques which can be difficult or labor intensive (such as resin transfer molding) in terms of resin infusion.

A more attractive way of producing effective, 3D laminates and reducing the impact weakness and delamination is a strategy known as "z-directional" toughening or "supplementary reinforcement" in which short fibres that align in the z-direction are introduced (perpendicular to the laminates) [24]. Early work by Garcia et al. [25] and Yamashita et al. [26] demonstrated this effect, predicting the need for fibres less than a micron in diameter, using silicon carbide whiskers of $0.1-0.5\,\mu m$ diameter. Low concentrations of filler led to improved edge delamination, although in-plane properties were also decreased. Jang and co-workers [27] reported work where whiskers of various types were incorporated into fibre composites, but these showed much less improvement than expected due to fibre clumping. The required concentrations also led to an increased viscosity and difficulty in handling and degassing materials, producing remnant voids. Nonetheless, Jang and other groups such as that of Sohn and Hu [28] showed that the use of short fibres such as Kevlar could lead to improved properties by mechanisms such as crack bridging if dispersion was sufficiently good. The concept of layered silicates as a potential supplementary filler for thermoset fiber composites will be introduced later in this review.

3
Crystallography and Surface Modification of Layered Silicates

Layered silicates belong to the structural group of swelling phyllosilicates minerals also known as 2 : 1 phyllosilicates or smectites. These minerals are often simply referred to as clays, with the term 'clay' by definition strictly referring to mineral sediments of particles with a dimension of less than $2\,\mu m$ [5]. The individual layered silicates are usually referred to by their mineral name (for example, montmorillonite) or rock name (bentonite) [5]. Montmorillonite is a rarely-found, neat silicate mineral and principal component of more common bentonite, which contains fine dispersions of quartz and other impurities [29]. Along with montmorillonite, commonly-used smectites include hectorite and saponite [30]. The main characteristic property of these layered minerals is their high aspect ratio and ability to swell via absorption of water and other organic molecules, leading to an increase in the interlayer distance.

Smectites consist of periodic stackings of approximately 1 nm thick layers. These layers form tactoids with thicknesses between $0.1-1\,\mu m$ [31]. The crystalline lattice of the silicate platelets consists of two tetrahedral silica sheets fused at the tip to a central octahedral sheet of alumina or magnesia [29].

Through sharing common oxygen atoms, as illustrated in Fig. 2, extended structures are formed [4]. Isomorphous replacement of central anions of lower valences in the tetrahedral or octahedral sheet results in negative charges on the silicate surface. Common substitutions are Si^{4+} for Al^{3+} in the tetrahedral lattice and Al^{3+} for Mg^{2+} in the octahedral sheet [5]. The negative charge on the platelet surface is counterbalanced by alkali or alkaline earth cations between the layers, known as the interlayer or gallery.

The number of sites of the isomorphous substitution determines the surface charge density and hence significantly influence the surface and colloidal properties of the layered silicate [32]. The charge per unit cell is thus a significant parameter necessary to describe phyllosilicates. The intermediate value for the charge per unit cell of smectites [33] ($x \approx 0.25$–0.6) compared to talc ($x \approx 0$) or mica ($x \approx 1$–2) enables cation exchange and gallery swelling for this group of phyllosilicates, making them suitable for epoxy nanocomposite formation [31]. The negative surface charge determines the cation exchange capacity, CEC [meq/100 g] which is key to the organic surface modification. The untreated smectite has a high affinity to water and thus does not readily absorb most organic substances including polymers, although some polymers such as poly(ethyleneoxide), poly(vinylpyrrolidone) and poly(vinyl alcohol) are able to access unmodified galleries. However, the low van-der-Waals forces between stacks do allow the intercalation and exchange of small molecules and ions in the galleries. In order to render the hydrophilic clay more organophilic, the inorganic ions in the gallery can be exchanged by the

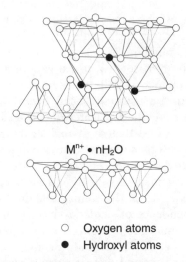

$M^{n+} \bullet nH_2O$

○   Oxygen atoms
●   Hydroxyl atoms

Fig. 2 Model structure of layered silicates (montmorillonite) where usually silicon sits in the tetrahedral locations of the oxygen network. The octahedral positions may variously be iron, aluminium, magnesium or lithium, and the exchangeable cation in the gallery is given by M^{n+} [4]

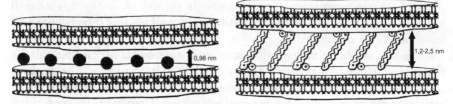

Fig. 3 Unmodified layered silicate (*left*) and layered silicate with interlayer-exchanged alkyl amine ions (*right*) [151]

cations of organic salts. Whilst the absorption of organic materials through cation exchange in montmorillonite has been the subject of studies for some years for various systems [32, 34], increasing detail on how the layered silicates can be rendered more accessible to epoxy resins has been reported [12, 35–38]. Fig. 3 illustrates the increase in layer spacing from less than 1 nm to 1.2–2.5 nm that occurs upon exchange with alkylamine ions. The degree of increased separation depends on the chemistry and length of the exchanged ions, as well as the charge density of the silicate.

4
Characterization of Thermosetting Layered Silicate Nanocomposite Morphology

The terms *intercalated, exfoliated* and *delaminated* are often used to describe the arrangement of the silicate platelets within the polymer matrix. Nanocomposite systems whose wide-angle diffraction spectra show no peaks in the diffraction angle range of $2\theta = 1$ to $6°$ are usually considered as effectively exfoliated. However, further investigations of the nanocomposite structure show that in many cases, the platelets are still arranged in regions of parallel platelets known as *tactoids*. It has been pointed out in the literature that the categories mentioned (intercalated, exfoliated) describe idealized morphologies only, and that most real structures fall between these extremes [39–41]. Vaia [42] thus suggested an expanded classification system to allow a more accurate description of a given layered silicate nanocomposite morphology. The expanded classification system considers aspects such as relative changes in d-spacing, the volume fraction of single platelets and aggregates and the dependence of single-layer separation on silicate volume fraction and critical volume fraction, and is shown in part in Fig. 4. Recent contributions by Morgan et al. [39, 43] and Kornmann et al. [44] also emphasise that both microstructure and nanostructure must be considered when fully describing a nanocomposite morphology. Since the techniques commonly applied to investigate such morphologies vary significantly in their

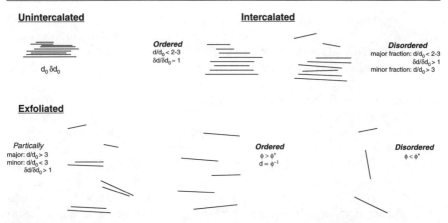

Fig. 4 Detailed nomenclature for the characterization of layered silicate nanocomposite structures. The arrangement of the layered silicates in the polymer matrix is classified on the basis of the relative change in d-spacing and correlation (d and δd); the relative volume fraction of layers and stacks of layers and the dependence of single-layer separation on the layered silicate volume fraction, ϕ [42]

resolution, as well as in the size of the area investigated, there is no single method that allows full description of any given morphology. Of all the techniques applied, the most commonly used are wide-angle X-ray diffraction (WAXD) and transmission electron microscopy (TEM). Most researchers recommend applying both as complementary tools to characterize and describe morphology. High resolution scanning electron microscopy (SEM) is increasingly becoming a tool that conveniently straddles a range of size-scales to describe dispersion, although clearly TEM is still required to investigate on the presence or otherwise of individual layers. Care must be taken though, if TEM is used, to not forget to seek and characterize larger-scale inhomogeneities.

4.1
Wide-angle X-ray diffraction

Of all the techniques used for the structure analysis of layered silicates and polymer nanocomposites, wide-angle X-ray diffraction (WAXD) is probably the most widely applied. The repeat distance between layers, the d-spacing, can be determined through the diffraction from two consecutive clay layer scattering planes. The distance between the two layer surfaces is known as the d-spacing. The two layers interact with the X-rays of the wavelength λ at the incident angle θ. A constructive interference occurs when the Bragg Law is fulfilled:

$$n \cdot \lambda = 2 \cdot d \cdot \sin \theta . \tag{1}$$

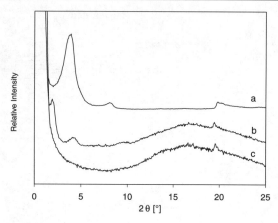

Fig. 5 XRD traces of (a) neat organoclay, (b) intercalated nanocomposite, (c) highly intercalated nanocomposite ($d_{(001)} > 90$ Å)

The integer n refers to the diffraction such that if $d_{001} = 1$ nm, then $d_{002} = 0.5$ nm. Therefore, with a known incident angle and wavelength, the layer
distance can be calculated. Figure 5 illustrates WAXD traces of intercalated
and exfoliated nanocomposite systems, compared to the spectra of pristine
clay. This technique can provide quick results with minimal sample preparation and allows direct determination of the average d-spacing between
silicate platelets. However, there are a number of limitations involved in this
method. Technically, the WAXD technique is often limited to a diffraction
angle of around $2\theta = 1°$ and hence (according to the Bragg Law), a maximum
d-spacing of 88 Å. The increase in the WAXD signal intensity at lower angles often makes it difficult to detect layer distances much above 65 Å [41].
The technique is highly dependent on the order of the clay, a distribution of
d-spacing and any disordered, non-parallel orientations which broaden and
weaken the WAXD spectra.

4.2
Small angle X-ray Diffraction (SAXD)

More recently examples of small angle X-ray scattering (SAXD) studies of
epoxy layered silicate nanocomposites and their in-situ formation have been
reported. Chin et al. [45] and Tolle and Anderson [46] have reported in-
situ SAXD studies on 1,4-diaminobenzene (MPDA) cured, diglycidyl ether of
bisphenol A (DGEBA)/octadecylamine montmorillonite systems using synchrotron radiation, as well as a standard small angle diffractometer [45], the
SAXD technique able to detect interlaminar spacings of up to 200 Å. Chen
and Curliss [47] recently presented a good example of synchrotron small
angle X-ray characterization of epoxy-based nanocomposites illustrated in
Fig. 6. The SAXD traces showed distinct peaks in the low angle regime, which

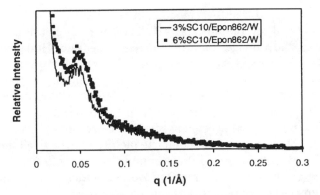

Fig. 6 Small-angle x-ray characterization of epoxy layered silicate nanocomposites synthesized using n-$C_{10}H_{21}NH_3^+$-montmorillonite (SC10), Shell Epi-Cure curing agent (W) and Shell Epon 862 resin where q is the scattering vector [47]

correlate with an interplanar spacings of 125 Å and 135 Å, respectively – all outside the range visible via WAXD.

4.3
Transmission electron microscopy (TEM)

Along with wide-angle X-ray diffraction, TEM is one of the most widely applied tools to investigate nanocomposite superstructure. With magnifications of up to 300 000 times, this powerful technique allows the edges of individual silicate platelets to be imaged. Figure 7 shows for example the TEM image of a diethyltoluenediamine (DETDA, ETHACURE® 100 of the Albemarle Cor-

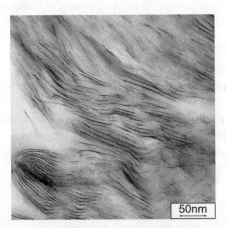

Fig. 7 TEM of DETDA cured DGEBA octadecyl ammonium modified layered silicate nanocomposite containing 7.5 wt % organoclay

poration, USA) cured DGEBA nanocomposite containing 7.5% octadecylammonium modified layered silicate [41]. TEM analysis is the main method that provides sharp images on a nanometer dimension. However, there are also a number of drawbacks involved with this method, in particular sample preparation by microtome cutting of ultrathin sections of only about 70 nm thickness is very labor intensive. Further, TEM only allows investigation of a very small area of a sample, problematic if the material is inhomogeneous. Ideally, a number of TEM specimens should be prepared from different sections of a sample to give a more complete picture of the overall morphology. Morgan et al. [39, 43] suggested that a combination of both, WAXD and TEM (at both low and high magnification) provides the most accurate representation of polymer-clay nanocomposite morphologies.

4.4
Optical and Scanning Electron Microscopy (SEM)

The use of optical or scanning electron microscopy has been reported in several instances to investigate the epoxy layered silicate morphology on a micrometer scale. Kornmann et al. [44, 48] compared the microstructure of various resin/layered silicate blends and their dispersions. Stacks of aggregates of layered silicates could be observed in such optical images. Salahuddin et al. [49] have used scanning electron microscopy to investigate the microstructure of highly filled epoxy nanocomposites containing up to 70 wt% layered silicate and the images clearly show a parallel alignment of the platelets. Generally, the technique of SEM (and optical microscopy) are very useful methods to analyze the distribution of the layered silicate on a larger scale. However, to investigate intercalation and exfoliation of the layered silicate higher resolutions such as provided by TEM and WAXD are required. As well as allowing the capture of the nanocomposite structure, SEM remains a particularly useful tool to investigate fracture surfaces of the nanocomposites.

4.5
Atomic Force Microscopy (AFM)

The use of the technique of atomic force microscopy to investigate the morphology of layered silicate nanocomposites has been rarely reported. Reichert et al. [50] investigated etched samples of thermoplastic poly(propylene)/clay nanocomposites using AFM in both the height and phase contrast mode. The images allowed detection of both large silicate platelets and finely dispersed silicates in skeleton-like superstructures. Zilg et al. [12] compared the interlayer distance of a nanocomposite determined from AFM and WAXD measurements. The d-spacing as determined from AFM images was 4.2 nm, larger than the value found using WAXD. Whilst this discrepancy could not

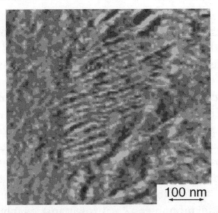

Fig. 8 AFM image of DETDA cure DGEBA nanocomposite containing 5 wt % octadecyl ammonium modified layered silicate [40]

be fully clarified, it was pointed out that the AFM tip may be able to deplete or strain the flexible silicate platelets in their more rigid epoxy surrounding. Figure 8 shows an AFM image of a diethyltoluenediamine cured DGEBA nanocomposite containing 5 wt % organoclay [40]. Although WAXD measurements of this material did not show any peaks, a structure of parallel, oriented platelets can be clearly observed.

4.6
NMR Dispersion Measurements of Nanocomposites

A rather new approach to determine the layered silicate dispersion in a polymer nanocomposite is through nuclear magnetic resonance (NMR) measurements [51, 52]. This method uses the reduction in the spin-spin relaxation time, T_1^H, of a nanocomposite when compared with the neat system, as an indicator for the organoclay layer separation. The work by VanderHart et al. [51] on polyamide-6 nanocomposites showed that the paramagnetic Fe^{3+} ions in the crystal lattice of the montmorillonite provide an additional relaxation mechanism of the protons. It is this additional relaxation which is determined by the average $Fe^{3+}-{}^1H$ distance (and therefore by Fe^{3+} ion and clay concentration) and its nano-dispersion throughout the polymer matrix, that determines the paramagnetic contribution to T_1^H. The paramagnetic contribution, (T_{1para}^H) was defined as:

$$\left(T_{1para}^H\right)^{-1} = \left(T_1^H\right)_{composite}^{-1} - \left(T_1^H\right)_{polymer}^{-1} , \qquad (2)$$

with $(T_1^H)_{composite}^{-1}$ being the inverse T_1^H of the composite and $(T_1^H)_{polymer}^{-1}$ the inverse T_1^H of the neat polymer.

5
Synthesis of Thermosetting Layered Silicate Nanocomposites

The main mechanism underpinning nanocomposite formation is that the monomer or polymer are able to intercalate into and react within (if the intercalant is monomeric) the interlayer galleries. Polymer nanocomposite formation can be divided into three primary classes, in-situ polymerization, intercalation of the polymer from solution, and melt intercalation of the polymer. Of these methods it is in-situ polymerization, which is relevant to the formation of epoxy-layered silicate nanocomposites. In-situ polymerization of layered silicate nanocomposite was first reported for the synthesis of nylon 6 polymer nanocomposites [53]. In this method, the organoclay is initially swollen by the liquid monomer (ε-caprolactam) enabling polymer formation outside and inside the interlayer galleries. The layered silicate gallery surface is pre-treated with 12-aminolauric acid which takes part with the ε-caprolactam in the reaction. For thermosets such as epoxy resins, a curing agent and heat are also required to promote the crosslinking reaction. There are a number of processing methods to produce such epoxy nanocomposites. In some of the early studies reported for rubbery epoxy systems, the layered silicate was added directly to the resin/hardener blend [37, 54, 55]. However, the more established methods of thermoset nanocomposite formation include pre-intercalation of the layered silicate by the resin for a period of time prior to addition of amine, and then subsequently reacted. Figure 9 shows a flowchart of the nanocomposite process as applied to the synthesis of most epoxy nanocomposites reported. Another processing techniques that have been examined include the use of a three-roll mill to impart high shear forces into the system [56]. In this work the clay is added to the epoxy which initially becomes more viscous and opaque, attaining clarity after shearing, subsequently mixed at higher temperatures with hardener.

Butzloff and D'Souza [57] investigated the controlled use of water in the synthesis of epoxy/alkylammonium modified montmorillonite systems.

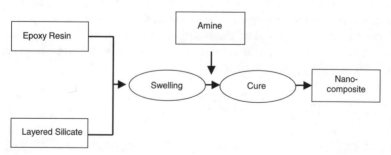

Fig. 9 Flowchart of the common process of thermosetting nanocomposite formation steps

The clay was treated with octadecylammonium and the flexible poly(oxy-propylene) diamines (Jeffamine series) of structure $NH_2CH(CH_3)CH_2 - [OCH_2CH(CH_3)]_xNH_2$, in this case Jeffamine D-230 (x is 2–3 on average). It appeared that the water led to bridging and increased agglomerate size, although this was subsequently reduced by ultrasonic treatment.

6
Controlling the Morphology of Epoxy Nanocomposites

6.1
Mechanism of clay dispersion

Several reports have discussed the mechanism of organoclay exfoliation during the in-situ polymerization of epoxy resins. Early work by Lan et al. [58] pointed out the important role of the balance between intergallery and extra-gallery reaction rates, as well as the accessibility of the resin and hardener monomers to the clay galleries during the exfoliation process. The common process for epoxy nanocomposite synthesis is to pre-intercalate the organoclay with the epoxy resin before cure at 50–100 °C for approximately one hour. It is reported that the monomers penetrate and swell the silicate layers until a thermodynamic equilibrium is reached between the polar resin molecules or resin/hardener blend, and the high surface energy of the silicate layers [48, 58]. The pre-intercalation leads to a limited increase in d-spacing [59]. The method of mixing *prior* to the cure process [60] or the additional use of solvents as processing aids [60, 61] was found to have little impact on the final nanocomposite structure. In fact, further increases in the distance between organoclay platelets require the driving force of the resin/hardener cure reaction or homopolymerization to overcome the attractive electric forces between the negative charge of the silicate layers and the counterbalancing cations in the galleries [58]. Decreasing polarity during reaction of the resin in the galleries displaces the equilibrium and encourages further monomer to diffuse into, and react within, the silicate galleries. It is found [41, 44] that shear forces from mixing *during* cure also improve the exfoliation process.

The change in interlayer spacing can be elegantly seen with *in situ* small angle X-ray scattering (Fig. 10) as the sample is heated up from 60–200 °C [62]. The clay originally has an interlayer spacing of some 18 Å, which increases to some 38 Å on mixing (seen at q of 0.16 Å^{-1}). From between 116–160 °C, a peak related to a spacing of 108 Å shifts to 140 Å, locating at about 150 Å at 200 °C. Chen et al. [63] divided the interlayer expansion mechanism into three stages. Stage one is the initial interlayer expansion due to resin and hardener intercalation of the silicate galleries. The second stage is

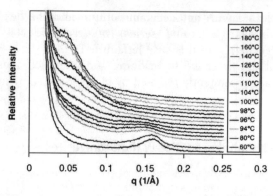

Fig. 10 In-situ small angle scattering showing disappearance of initial *d*-spacing, and appearance of new exfoliated material at small angle (large *d*-spacing for 3% clay) for an organo-ion exchanged montmorillonite (surface treatment is octadecyl ammonium), with a bisphenol F epoxy [47]

the interlayer expansion state where the interlayer spacing steadily increases due to intergallery polymerization. The third stage of interlayer expansion is characterized by a decreased interlayer expansion rate. In some cases, a slight decrease in interlayer spacing could be observed before cessation of gallery change, due to restrictions on further extragallery change because of gelation.

Recent work by Kong and Park [64] likewise defined an exfoliation process which occurs in three distinct steps or stages (rather than a gradual process) for the isothermal cure of DGEBA with 4, 4′-diaminodiphenyl sulphone (DDS) and an octadecylammonium-treated montmorillonite. Figure 11

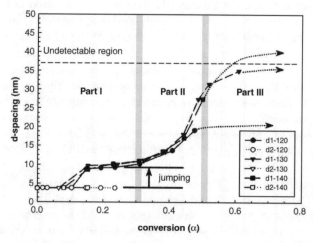

Fig. 11 Increase in *d*-spacing as a function of conversion for the isothermal cure of a DGEBA/DDS/octadecylamine montmorillonite at various temperatures [64]

shows the three different stages as a function of conversion. The stages are: 1st stage – ingress of the DGEBA monomer, 2nd stage – the self polymerization of the DGEBA due to the catalytic effect of the organo-ion and 3rd stage – the crosslinking of the epoxy residing in the gallery with the amine, in the presence of the amine (note: DDS was largely used in preference to 4, 4'-diaminodiphenylmethane, DDM) due to the lower reactivity of DDS.

Jiankun et al. [60] outlined a general thermodynamic approach to the exfoliation process. According to their work, the ability of a resin system to exfoliate a layered silicate is determined by the change in Gibb's free energy during cure, exfoliation occurring if $\Delta G = \Delta H - T\Delta S < 0$. It is assumed that the change in entropy is low according to previous reports [65, 66] and can thus be disregarded. Hence the exothermal curing heat of the intergallery epoxy resin, ΔH_1, must be higher than the endothermic heat to overcome the attractive forces between the silicate layers, ΔH_2. If $\Delta H_1 > \Delta H_2$, the clay exfoliates during cure.

Very recent work by Park and Jana [67] challenges that contention the intra- vs. extra-gallery reaction rate is the only key aspect that determines degree of intercalation. They found that a faster intra-gallery reaction rate accelerated (but did not necessarily enhance) the degree of exfoliation. Experiments were performed in which the two intra- and extra-gallery rates were carried out by mixing the same organic modifier residing in the clay galleries, into the resin. Despite matched intra- and extra-gallery rates, exfoliated structures were still achieved. Their view is that properties such as storage modulus and viscosity are the most crucial determinants of the final degree of delamination, being reflections of issues such as attraction between platelets and elastic recoil energy. The attractive forces are between the clay and organically-modified ions and between the organo-ions themselves. Because of these forces, the epoxy monomer that becomes crosslinked cannot relax and the elastic forces increase until they overcome the attractive, interlayer forces. The ability of the growing network to push apart the silicate layers depends also on the viscosity of the medium through which the nanoplatelets must move, gelation ultimately causing cessation of any such motion. Indeed, an empirical conclusion of their work is that complete exfoliation is encouraged if the ratio of shear modulus-to-complex viscosity is greater than $2-4\,s^{-1}$ during the cure process. Following from this, is the concept that the outer layers are able to separate first, and that this occurs more readily at the periphery of the tactoids where they are less constrained. Higher cure temperatures favour exfoliation because the elastic buildup is greater and viscosity (initially) is lower, so the plates are able to separate before viscosity eventually rapidly increases and gelation causes cessation of further motion. These ideas provide a more sophisticated understanding of the intercalation/exfoliation mechanism, implying that even though a faster intragallery polymerization enhances the rate of intercalation/exfoliation, it is not the only important factor.

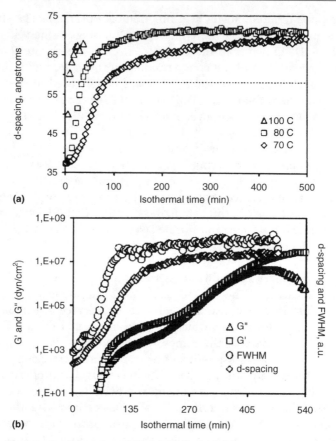

Fig. 12a,b Changes in properties of isothermally cured epoxy monomer and montmorillonite surface-treated with methyl, tallow, bis-2-hydroxyethyl ammonium cations at
10 wt % clay:(**a**) *d*-spacing at a range of temperatures and (**b**) oscillatory rheological parameters and *d*-spacing at 70 °C [63]

Chen et al. [63] also presented a number of rheological parameters at
isothermal temperature, measured as a function of cure. They determined
the change in interlayer expansion with time, as shown in Fig. 12a and found
an increase with reaction. In some instances, especially at low temperatures
and compositions not shown here, they found a slight decrease in gallery
spacing at high conversions, ascribed to the curing system compressing the
layers slightly. Their isothermal rheological data (Fig. 12b) showed that storage modulus increased with increasing *d*-spacing but this ceased when the
rubber-to-glass transition occurred. It is hypothesized that the interlayer expansion ceases when the modulus of the material outside the gallery became
equal to that within it, and that if the outer modulus increased further, the
compression mentioned above was a result.

In comparison to the formation of thermoplastic nanocomposites, the transformation from the liquid reactive resin to the crosslinking solid strictly limits the exfoliation of clay in thermosetting nanocomposites to a small processing window. It was found that significant changes in the interlayer distance occur at the early stage of cure, before gelation restricts the mobility of the clay platelets [68]. The aspects discussed in the next sections have been found to influence the degree of epoxy nanocomposite formation and resultant morphologies.

6.2
The Nature of the Silicate and the Interlayer Exchanged Ion

Two fundamental aspects of the organoclay determine the formation of epoxy nanocomposites from in-situ polymerization: the ability of the interlayer exchanged ion to act as a compatibilizer and render the layered silicate 'epoxyphilic', and the catalyzing effect of the exchanged ion on the polymerization reaction in the galleries [31].

The charge density of the smectite determines the concentration of ions in the interlayer galleries that can be exchanged, and therefore the amount of epoxy monomer that can be preloaded in the galleries of the modified organoclay. Lan and coworkers [58] investigated the effect of various smectites with cation exchange capacities (CEC) ranging from 67 mmol equivalent/100 g (meq/100 g) for hectorite. to 200 meq/100 g for vermiculite. They found that silicates with intermediate layer charge densities such as montmorillonite and hectorite, are well suited for layered silicate modification. Generally, layered silicates with a low charge density are more readily accessible for intragallery polymerization than high charge density clays with a higher population of gallery onium ions, the low charge density silicates thus yielding greater degrees of exfoliation. Kornmann et al. [48] reported similar results for two different montmorillonite clays, with a CEC of 94 and 140 meq/100 g. After modifying the layered surfaces of the clay with an octadecylamine ion, TEM images of the DGEBA/poly(oxypropylene) diamines (Jeffamine D-230) nanocomposites showed regular stacks of 9 nm for the high CEC, and 11 nm for the low CEC montmorillonite. The difference in organoclay layer separation was assumed to be due to the population density of alkylamine ions in the galleries and hence the space available into which the epoxy can ingress.

The nature of the interlayer exchanged ion also significantly determines the compatibility of the layered silicate with the epoxy resin, as well as the inter-gallery reaction rate. Pinnavaia and coworkers performed several studies using different types of organically modified layered silicates [55, 58, 69–73]. Before the intergallery reaction can be initiated, the clay tactoids must be preloaded with the epoxy monomer. The population density of the gallery onium ions and the basal spacing of the smectite determine the initial accessibility of the epoxy and hardener monomers to the clay galleries.

It was shown that for a series of alkylammonium ion, $CH_3(CH_2)_{n-1}NH_3^+$, exchanged montmorillonite with $n = 4, 8, 10, 12, 16$ and 18, the length of the alkylammonium ion greatly affects clay expansion before cure [58]. The degree of exfoliation during cure was also found to depend on the amount of pre-intercalated resin in the clay galleries, an alkylamine cation chain length with greater than eight methylene groups was found to be necessary for nanocomposite formation. In more recent work by Zilg et al. [74], it was reported that the alkyl chain length for an organically modified fluorhectorite had to exceed six carbon units to promote intercalation or exfoliation. A chain length higher than eight units did not further improve the exfoliation process.

Wang and Pinnavaia [37] found in a study on a series of primary to quaternary octadecylammonium ion modified clays illustrated in Fig. 13, that primary and secondary onium-treated clays became exfoliated in a DGEBA/1,4-diaminobenzene (MPDA) system, whilst the tertiary and quaternary ion-modified clays remained intercalated. This effect has been ascribed to the higher acidity and stronger Brönsted-acid catalytic effect of the primary and secondary onium ions on the intergallery epoxy reaction. Since the tertiary and quaternary ions are less acidic, intercalation of the curing agent is less favorable due to changes in the gallery expansion of these nanocomposites. These results are in good agreement with more recent work reported by Zilg et al. [12] who reported that nanocomposite formation based upon various layered silicate modification has shown that ion exchange with protonated primary amines, such as 1-aminodedecane ions, gave larger interlayer distances in the nanocomposite than those based on quaternary amine modification (for example N,N,N-trimethyldodecylamine ions).

Messersmith and Giannelis [75] investigated the formation of a quaternary bis (2-hydroxyethyl) methyl tallow alkylammonium ion modified montmorillonite DGEBA nanocomposite using three different curing agents (nadic methyl anhydride (NMA), benzyldimethylamine (BDMA) and boron trifluoride monoethylamine (BTFA)). The tallow unit is widely used as the organic component in organo-ions and are alkyl tails, usually hydrogenated, obtained from naturally occurring oils. According to the Southern Clay Products mate-

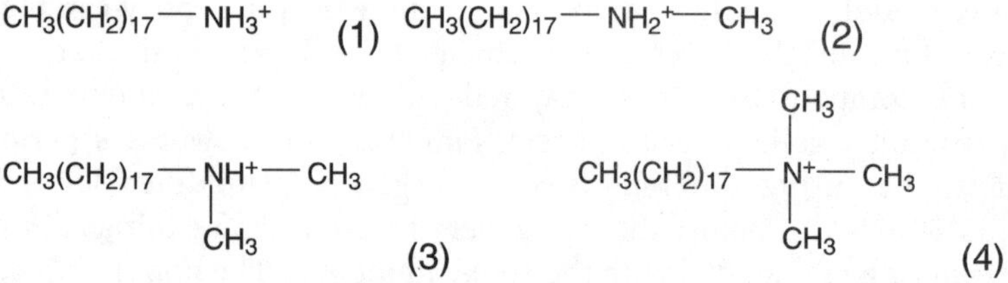

$$CH_3(CH_2)_{17} - NH_3^+ \quad (1) \qquad CH_3(CH_2)_{17} - NH_2^+ - CH_3 \quad (2)$$

$$CH_3(CH_2)_{17} - \underset{\underset{CH_3}{|}}{NH^+} - CH_3 \quad (3) \qquad CH_3(CH_2)_{17} - \underset{\underset{CH_3}{\overset{CH_3}{|}}}{N^+} - CH_3 \quad (4)$$

Fig. 13 Examples of primary (1), secondary (2), tertiary (3) and quaternary (4) octadecyl amine ions – the more acidic ions (1), (2) favour exfoliation

rials information, they usually consist of a mixture of different length chains, $\sim 65\%$ C_{18}; $\sim 30\%$ C_{16}; $\sim 5\%$ C_{14}. For this particular modified clay it was found that bifunctional primary and secondary amines resulted in immediate clouding of the resin, with little or no increase in layer separation. It was assumed that this behaviour was due to bridging of the silicate layers by the bifunctional amine which prevented further layer expansion. However, primary and secondary amines were both found to be effective for exfoliation of the alkyl ammonium ion exchanged clays. An interesting reported method of clay modification [76] involved the use of poly(oxypropylene) diamines (Jeffamines) themselves as modifiers, as they are also commonly used as epoxy curing agents. Unmodified sodium montmorillonite is added to the appropriate-length poly(oxypropylene) diamine (Jeffamine D-2000, where x is on average 33 was found to be the best) with hydrochloric acid used to form the quaternary ammonium salt. This treated clay is subsequently added to a Jeffamine/epoxy mixture and cured. Good levels of intercalation and exfoliation are seen, with a three-fold increase in modulus and improved thermal and solvent resistance.

Long primary linear chain alkylammonium ions such as $CH_3(CH_2)_{17}NH_3^+$ (octadecyl ammonium) have proven to be the most appropriate organo-ions for the synthesis of exfoliated systems. In recent times the bis (2-hydroxyethyl) methyl tallow alkylammonium organo ion has also been widely used, shown in Fig. 14 [77]. Other workers have experimented recently with other modifications to these organo-ions, such as Feng et al. [78], who modified the afore-mentioned bis (2-hydroxyethyl) methyl tallow alkylammonium ion with tolylene 2,4-diisocyanate and bisphenol A, forming an organo-ion much greater in length but still hydroxy-terminated and containing polar, internal amine and carbonyl functionalities. However, this was only partially successful in the sense that the basic treated clay increased some 13 Å from that of untreated material, but the final d-spacing in the cured composite only increased a further 7 Å. Nonetheless, rubbery moduli were enhanced, as was the value of the glass transition temperature.

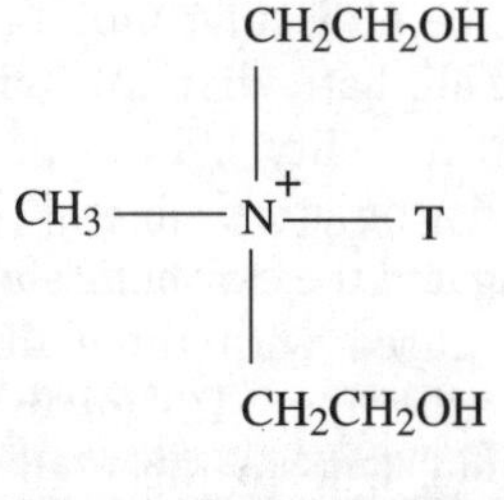

$$CH_3 \longrightarrow \overset{+}{\underset{|}{\overset{|}{N}}} \longrightarrow T$$

CH$_2$CH$_2$OH (top)

CH$_2$CH$_2$OH (bottom)

T is tallow (ca. 65% C18, 30%C16, 5% C14)

Fig. 14 Structure of methyl, tallow, bis-2-hydroxyethyl methyl tallow ammonium ion commonly used in epoxy nanocomposites

6.3
Curing agent

Although the silicate interlayer exchanged ions have been widely studied with respect to control of intercalation or exfoliation of the nanocomposite system, the choice of a suitable curing agent is also reported to be a significant factor determining delamination of the thermosetting nanocomposite system. Recent research by Jiankun et al. [60] using montmorillonites modified with $CH_3(CH_2)_{17}NH_3^+$ and $CH_3(CH_2)_{17}N(CH_3)_3^+$ ions has shown that for DGEBA-based nanocomposites, low viscosity curing agents (methyltetrahydrophthalic anhydride – MTHPA) intercalate more easily into the clay galleries than the highly viscous curing agent, 4, 4'-diaminodiphenylmethane. It should be noted that this hardener is a solid during the first processing stage (initial mixing of the nanocomposite) during which time mass transfer, and thus intercalation, of the hardener into the clay galleries is difficult.

Kornmann et al. [79] investigated the correlation of diffusion rate and reactivity of a DGEBA system and the subsequent degree of exfoliation. It was shown that the molecular mobility and reactivity of the curing agent are important factors affecting the balance between intergallery and extragallery reaction. For the three different curing agents investigated, the poly(oxypropylene) diamines (Jeffamine D-230), 3, 3'-dimethyl4, 4'-diaminodicyclohexylmethane (3DCM) and 4, 4'-diaminodicyclohexylmethane (PACM), it was found that the Jeffamine D-230 gave better exfoliation at comparable degrees of conversion of a DGEBA/octadecylammonium montmorillonite system, than the cycloaliphatic polyamines of higher reactivity (3DCM and PACM) after three hours of cure at 75 °C. However, an attempt to improve exfoliation of these systems by reducing the cure temperature and reactivity was not successful. The exfoliation process was thus assumed to also depend on other factors such as diffusion rate of the amine into the clay galleries. Solubility parameters of the various components (determined from the group contribution method) were used as an indication of their polarities, and were found to be in the same order of magnitude for the three different amines. It was thus assumed that molecular flexibility is also a determining factor for the molecular mobility or diffusion rate. This is based on observation that the aliphatic diamine, with its highly flexible backbone and improved molecular mobility, led to significantly better exfoliation when compared with the rigid aromatic amines.

Kong and Park [64] investigated the exfoliation behaviour of DGEBA/octadecylamine-montmorillonite cured with three different high-performance curing agents: m-phenylenediamine, 4, 4'-diaminodiphenylmethane and 4, 4'-diaminodiphenyl sulphone. Due to the high melting temperature, the DDS-cured system was synthesized at 150 °C, rather than 75 °C. Since changes in cure temperature may affect a number of parameters such as catalytic effects and reaction rate, comparison between these systems should be undertaken with caution. However, some interesting results can be found for the two resin systems (different amines, MPDA, DDM) cured at the same temperature. It was found that the DDM gave better exfoliation than the MPDA hardener. Improvements in exfoliation were related to the reactivity of the amine, as indicated by their electronegativities, with lower reactivities (electronegativity) leading to better exfoliation.

Chen and Curliss [80] have reported an interesting way of making epoxy nanocomposites by the use of electron beam (e-beam) curing and cationic polymerisation. This process has advantages over conventional methods, some being: low volatiles, low energy usage and lower temperatures of reaction – whilst maintaining the good properties of the epoxies, such as low shrinkage. They found an increase in d-spacing of an octadecylammonium treated clay of some 27 Å, compared to over 100 Å from thermal cure using the same treated clay. This lesser increase was largely due to the rapid rate of reaction caused by e-beams. It was found that the modulus of the epoxy nanocomposite could be increased by this method, without the decrease in glass transition temperature seen in thermally cured systems. Since e-beam processing is increasingly being used in multi-phase systems such as carbon fiber composites, it could become a useful nanocomposite processing tool in the future.

6.4
Cure Conditions

The effect of the cure temperature on nanocomposite formation has been the subject of several studies. In most cases [41, 44, 46, 79] it was found that higher cure temperatures gave better exfoliation of the organosilicate in the epoxy matrix. This improved exfoliation was mainly ascribed to the higher molecular mobility and diffusion rate of the resin and hardener into the clay galleries, leading to an improved balance between inter- and extra-gallery reaction rate. Recent *in situ* small angle WAXD studies on a synchrotron by Tolle and Anderson [46] have shown quantitatively that for m-phenylenediamine cured octadecylammonium montmorillonite/DGEBA nanocomposites, increased cure temperatures caused organoclay exfoliation in a shorter period of time, and increased the magnitude of the basal spacing in the final morphology. Work on three different high-performance resin systems, based on three different epoxy resins (diglycidyl ether of bisphenol A – DGEBA, triglycidyl p-amino phenol – TGAP and tetraglycidyl ether of 4, 4'-diaminodiphenylmethane – TGDDM) and the low viscosity, aromatic hardener diethyltoluenediamene hardener found that increased cure temperatures improved exfoliation [41]. Lan et al. [81] also found improved exfoliation when the moulds were preheated to the cure temperature prior to filling the prepolymer into the modulus.

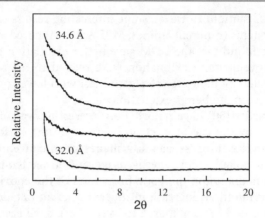

Fig. 15 X-ray traces of epoxy nanocomposites – intermediate cure temperaturesshow best exfoliation (the cure cycles are indicated above the traces) [58]

It appears from the literature that the effect of cure temperature on exfoliation of layered silicates in the epoxy matrix varies between systems. Whilst the exfoliation process of some epoxy systems are found to be independent of the cure temperature [44, 60], Lan et al. [58] reported an optimum temperature interval in one system in terms of organoclay delamination, higher or lower cure temperatures being disadvantageous. Figure 15 shows WAXD traces of m-phenylenediamine cured DGEBA/$CH_3(CH_2)_{15}NH_3^+$-montmorillonite nanocomposites cured at different temperatures. In that work it was theorized that too low cure temperatures may lead to intercalation rates that are slow, and if extra-gallery polymerization dominates intercalated (rather than exfoliated) structures will dominate. Cure temperatures that are too high were conversely thought to favor extragallery polymerization [58]. Thus, depending on the nature of the resin and curing agent, cure cycles should be designed to balance the intra- and extragallery polymerization rates. It has been claimed that cure should preferably involve exfoliation at lower temperatures, with a subsequent cure at elevated temperatures [58, 71, 82] since rapid cure too early in the reaction may lead to encapsulated tactoids.

6.5
Other Strategies for Improved Exfoliation

A number of other strategies to manipulate epoxy nanocomposite formation have been discussed in the literature. Chin et al. [45] have investigated the influence of the stoichiometric resin/hardener ratio on exfoliation of a MPDA/DGEBA/octadecylammonium montmorillonite using *in situ* small-angle X-ray scattering. It was found that resin cure with under-stoichiometric amounts of MPDA and the homopolymerization of DGEBA without any hardener (as earlier reported by Lan et al. [71]) leads to the formation of exfoliated nanocomposites. Further, it was found that the exfoliation was improved with decreasing amine concentrations. The extragallery reaction dominated for stoichiometric resin/hardener ratios and greater, due to increased rates of extragallery reaction and thus intercalated structures.

The use of low-boiling solvents such as acetone to enhance the process-ability, and hence the final nanocomposite structure has been investigated by Brown et al. [61]. Their work has shown that the preloading of the layered silicate with the resin could be processed at significantly lower temperatures due to the decreased viscosity compared to the neat systems and no change in the curing reaction, final morphology or mechanical properties was observed. More recently, Salahuddin et al. [49] synthesized highly-filled epoxy organoclay nanocomposite films of up to 70% montmorillonite using acetone as a processing aid. The low boiling solvent was necessary to enable mixing of such high clay levels with the resin/hardener blend. The final material was a transparent film with the clay platelets being arranged with d-spacings of 30 to 70 Å.

Triantafillidis et al. [83, 84] recently investigated a new approach of epoxy layered silicate nanocomposites with organic modifier. In their work, the layered silicate was treated with diprotonated forms of poly(oxypropylene) diamines (Jeffamines) of the ionic form α, ω-[NH$_3$CHCH$_3$CH$_2$(OCH$_2$CHCH3)$_x$ NH$_3$]$^{2+}$ (with x approximately 2–3, 5–6, and 33). Some of the inorganic, intergallery ions were only partly exchanged, using less than stoichiometric ratios of organic modifier. The silicate modifier not only plays the role of a surface modifier and polymerization catalyst, but also of the curing agent. This strategy improved mechanical properties and greatly reduced the plasticizing effect of the modifier that can often be found with higher levels of ion-exchange in the more commonly used mono-amine modified layered silicates.

7
Properties of Thermosetting Nanocomposites

7.1
Cure Properties

Knowledge about the curing behaviour, in particular about the thermal transitions such as gelation and vitrification during cure of the epoxy system, are of vital importance in the optimization processing conditions and the final properties of the crosslinked polymer [85–88]. To date, the effect of unmodified [89] and organically-modified layered silicates [61, 68, 71, 90, 91] on epoxy cure kinetics has mainly been investigated by differential scanning calorimetry (DSC). Bajaj et al. [89] investigated the effect of unmodified mica on the curing behaviour of a 4, 4′-diaminodiphenylmethane-cured diglycidyl ether of bisphenol A resin. It was found that mica accelerates the curing reaction substantially. In their work it was proposed that the hydroxyl-groups on the mica surface act as hydrogen bond moieties that accelerate the crosslinking reaction through participation in the glycidyl-ring opening process illustrated in Fig. 16. The group of Pinnavaia et al. [71, 92] also found that acidic onium ions catalyze self-polymerisation of DGEBA at increased temperatures, as judged by DSC. The mechanism for the homopolymerization in the organoclay galleries was proposed, and is shown in Fig. 17, where protons are formed through dissociation of the primary alkylammonium cations attacking the glycidyl-ring and thus catalyse homopolymerization.

Brown et al. [61] have used DSC to investigate the influence of two different organoclays on the homopolymerization of neat DGEBA and on the reaction of a DGEBA cured with poly(oxypropylene) diamines (Jeffamine D-2000, $x = 33$). It was found that both the epoxy homopolymerization and the amine-cured reaction were mildly catalyzed through the presence of a dimethyl ditallow ammonium montmorillonite (Rheox B34) where the

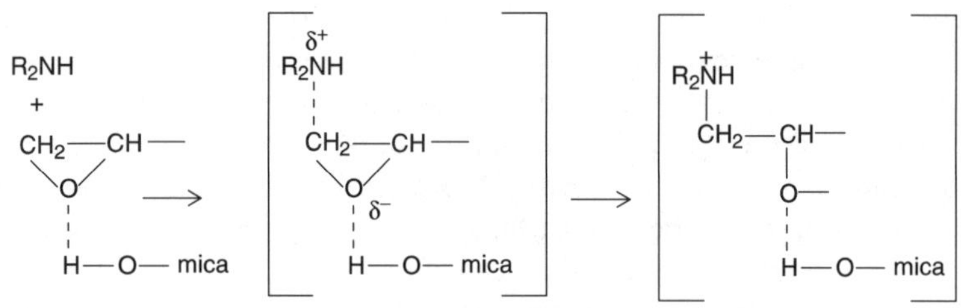

Fig. 16 Proposed catalytic effect of unmodified mica on epoxy [89]

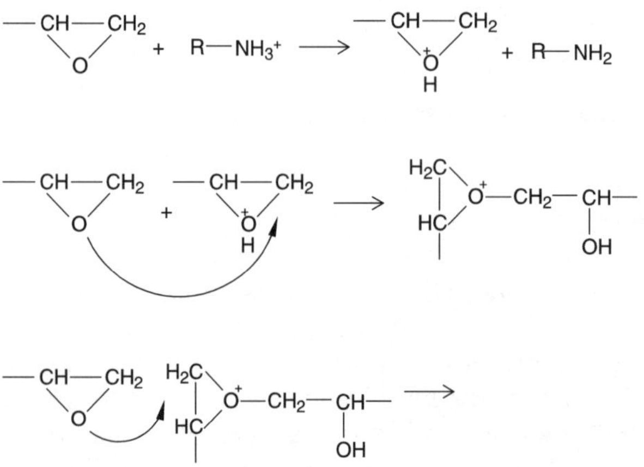

Fig. 17 Proposed ctalytic effect of the organoclay on epoxy homopolymerization [58]

organoclay showed only a slight increase in the interlayer distance during cure. More significant catalytic effects were observed for a bis(2-hydroxy-ethyl)methyl tallow ammonium montmorillonite (Cloisite 30A®, Southern Clay Products), where the organoclay exfoliated during the resin/hardener cure reaction. Recent work by Xu et al. [91] investigated the cure behaviour of diethylenetriamine/DGEBA/$CH_3(CH_2)_{15}N(CH_3)_3$–montmorillonite. In this work a slightly decreased activation energy was found with increasing organoclay concentration.

Ke et al. [68] reported decreasing gelation times with increasing organoclay concentration for a N,N-dimethylbenzylamine (DMBA) cured DGEBA containing 0–7 wt % organoclay, using a process where the resin is stirred on a heated plate until the resin can be pulled into continuous fibres. However, as this work was focused on other aspects, no explanation for the reduced gelation time was given. Further reaction kinetics studies on a poly(oxypropylene) diamines (Jeffamine D-230, $x = 2$–3) cured DGEBA system using an octadecylammonium modified montmorillonite were reported by Butzloff et al. [90], where the kinetics of layered silicate/epoxy resin and layered silicate/hardener/epoxy resin were investigated using DSC. For the two-part mixture, a significant decrease in enthalpy was reported for modified clay concentrations greater than 5 wt %. Interestingly, the three-part mixture showed a maximum in activation energy at 2.5 wt % organoclay concentration. A composition dependence on exfoliation was also reported with mixtures of intercalated and exfoliated layered silicates for concentrations above 2.5 wt %.

A recent study on the influence of an octadecylammonium-modified montmorillonite on the crosslinking reaction of different diethyltoluene diamine cured epoxy resins, showed a decrease in gelation time due to the catalytic effect of the organo-ions on resin cure [93]. This study compared the different techniques of DSC, chemorheology and dynamic mechanical thermal analysis (DMTA) for monitoring cure. The latter is a technique known as the 'flexural braid' test and is based on techniques developed by Gillham and Aronhime [94, 95]. It was found that the cure kinetics of the DGEBA/DETDA resin systems is more strongly affected by the addition of organoclay than the other two systems, based on the TGAP and TGDDM resins. This is likely due to the fact that the organoclay exfoliates better in the DGEBA, exposing more treated surface area to catalyze homopolymerization and resin hardener cure. Both gelation and vitrification times of the resin systems steadily decreased with increasing filler concentration. The decreased gelation time was found to be solely to the increased rate of reaction, rather than the formation of a physical gel (at lower conversion), as can often be observed in layered silicate dispersions. The actual degree of conversion at gelation was little changed.

Although most epoxy nanocomposite research work is based on amines as the curative, other hardeners are increasingly being reported. One system being increasingly reported involves imidazoles, which uses anionic catalysis to initiate homopolymerisation. This occurs by converting epoxy to hydroxyl units, which can then react further with another epoxy unit. It is a complex curing system with adduct formation, etherification and imidazole generation being some of the key steps [96]. This complexity shows up in a variety of behaviours in different studies to date. It was reported that in one such system, the addition of clay decreased the rate of cure, particularly at high cure temperatures [97]. In this case, the reaction mechanism was found to be little changed, the organically-treated montmorillonite largely thought to be topologically obstructive with regards to diffusion of resin and reaction within the gallery. However in other epoxy – nanoclay work, imidazoles decreased the time to gelation upon clay addition [98–100]. Most of the work to date has fitted only general reaction models to the imidazole cure data and no effort has been made to follow more closely the effect on various sub-processes of the epoxy-imidazole curing reaction, which may in part explain the differing results for different imidazoles. Epoxy nanocomposites of anhydride-cured materials have also been reported with the addition of small amounts of imadazole as the catalyst [101] where the treated clay was found to strongly affect the reaction kinetics of the epoxy/anhydride system.

7.2
Thermal Relaxations

The effect of the organoclay on the α-relaxation or glass transition temperature (T_g) has been the subject of a number of studies. In some cases a constant or increased T_g has been reported with increasing organoclay addition [61, 72, 75, 102–104]. Increased T_gs were also reported for a series of vinyl ester nanocomposites [105] and cyanate ester-layered silicate nanocomposites [106]. A peak broadening and increase in T_g has been related to restricted segmental motions near the interface between the organic and inorganic phase [75]. Kelly et al. [103] investigated DGEBA-layered silicate nanocomposites cured with Epon "R" V-40 (Henkel), a condensation product of polyamines with dimer acids and fatty acids, and found an increase in T_g when the organoclay was initially swollen with the curing agent, rather than the epoxide. It was concluded that initial swelling of the layered silicate in the curing agent leads to better epoxy absorption.

Others have found a reduction in the glass transition temperature with increasing organosilicate content. It was reported that highly crosslinked high glass transition temperature resin systems [40, 44] led to a steady decrease in T_g with increasing organo-ion concentration. Figure 18 shows the reduction in T_g of various glassy epoxy nanocomposite systems with increasing organoclay concentration [40]. The complexity of the cure reaction and possible side reactions involved made it difficult to determine the governing factor causing the reduction in T_g. The crosslink density in the filled systems may be

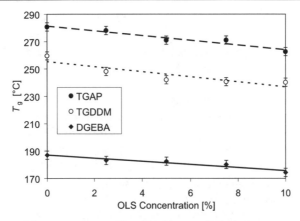

Fig. 18 Glass transition temperature, T_g, of highly crosslinked epoxy nanocomposite systems. The resin systems are diethyltoluene diamine cured octadecylammonium modified layered silicate DGEBA, TGAP and TGDDM [63]

decreased or the matrix plasticised by smaller molecules present in the network [44]. The organoclay may catalyze epoxy etherification and unreacted entrapped resin, hardener or compatibilizer molecules may act as a plasticizer. In addition the high cure temperatures required for these resin systems may degrade the layered silicate surface modifier, which are nominally stable to about 200–250 °C.

Chen et al. [63] found a decreased T_g for an hexahydro-4-methylphtalic anhydride cured epoxy (3,4-epoxycyclohexylmethyl-3,4-epoxycyclohexane) layered silicate nanocomposite. The layered silicate was rendered organophilic through bis (2-hydroxyethyl) methyl tallow alkylammonium cations. The decrease in T_g was proposed to be due to the formation of an interphase consisting of the epoxy resin, which is plasticized by the surfactant chains. Triantafillidis et al. [83] reported that limiting the clay modification reduces the plasticizing effect due to the organic modifier.

In contrast, recent work on intercalated phenolic-based cyanate ester nanocomposites by Ganguli et al. [106] has shown a significant increase in T_g through organoclay addition: As determined from the onset in storage modulus of dynamic mechanical analysis, the nanocomposite containing 5% organoclay showed a T_g of 390 °C compared to a T_g of 305 °C for the neat material. Whilst possible reasons for this significant increase were not discussed in this paper, it was pointed out that the loss modulus traces of the cyanate ester nanocomposites exhibited multiple, broad relaxations which is indicative of a heterogeneous cross-link topology.

A recent multiple-relaxation model by Lu and Nutt [107] has been proposed and is illustrated in Fig. 19, to explain the range of behaviours that have been observed in terms of the glass transition of epoxy materials. It divides up the nanocomposite into 3 domains: Domain I – a slow relaxation due to teth-

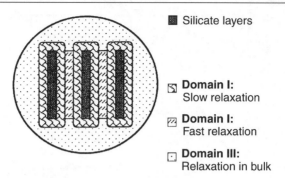

Silicate layers

Domain I:
Slow relaxation

Domain I:
Fast relaxation

Domain III:
Relaxation in bulk

Fig. 19 Model proposed for polymer relaxation in epoxy nanocomposites, showing regions of various relaxation rates [107]

ering of the epoxy molecules to the organo-ions (particularly if they are reactive, such as with terminal hydroxyl units), Domain II – a faster relaxation due to what could be called the "isolation" effect where lack of entanglement of chains and reduced cooperativity mean that confined and isolated polymers experience more rapid motions. This increased mobility of isolated polymer chains has also been observed in thermoplastic nanocomposites such as polystyrene, as judged by NMR spectroscopy [108]. Domain III – the bulk relaxation. Clearly in intercalated systems, there is an enhancement of the fast relaxation (Domain II) but this decreases as the system becomes more exfoliated and the main effect becomes the tethering of the epoxy chains to the surface of the clay. Of course, this latter aspect depends on the degree and strength of the tethering and interaction between epoxy and clay/organo-ion, as well as the volume of material which interacts in such a way with the clay. This raises the concept of a form of "percolation" of the clay in terms of the amount of the epoxy matrix which is influenced and predicts a strong effect on the glass transition due to clay concentration.

To date only few dielectric relaxation studies have been reported on thermosetting nanocomposite systems. Kanapitsas et al. [109] reported isothermal dielectric relaxation studies of epoxy nanocomposite systems based upon three different clay modifications, a low viscosity epoxy resin based on the diglycidyl ether of bisphenol-A type (Araldite LY556, CIBA) and an amine hardener in a temperature range of 30–140 °C. Whilst details on the epoxy system investigated and the nanocomposite morphology were vague, it was reported that the overall mobility is reduced in the nanocomposite compared to the neat matrix resin.

Little study has been made on the effect of layered silicates on secondary relaxations. Recent work [40] has shown a decrease in the (sub-ambient) β-relaxation temperature in the order of 5–7 °C from initially -50 °C to -58 °C towards -56 °C to -62 °C due to the addition of 10% layered silicate, demonstrating that the presence of layered silicate also affects the mobility of epoxy sub-units within the glassy state.

7.3
Mechanical Properties

7.3.1
Flexural, Tensile and Compressive Properties

In early work by Messersmith and Giannelis [75] on epoxy systems, a nadic methyl anhydride-cured DGEBA-based nanocomposite containing 4 vol % silicate showed an increase in the glassy modulus by 58% and a much greater increase of some 450% in the rubbery region. Pinnavaia and coworkers [37, 54, 55, 58, 70] investigated a number of intercalated and exfo-liated rubbery and glassy epoxy nanocomposites. A series of nanocomposites based on DGEBA, poly(oxypropylene) diamines (Jeffamine D-2000, $x = 33$) and a range of $CH_3(CH_2)_{n-1}NH_3^+$ montmorillonites with $n = 8$, 12, 18, showed a steady increase in both tensile strength and modulus with in-creasing chain length and organo-clay concentration. More than a 10-fold increase in strength and modulus was achieved through addition of 15% of the $CH_3(CH_2)_{17}NH_3^+$ modified montmorillonite. The degree of reinforce-ment was found to be dependant on the extent of exfoliation. It is assumed that the alignment of platelet particles under strain contributes to the signifi-cant improvement in the rubbery state. This alignment enables the platelets to function like long fibres in a fibre reinforced composite [55]. Rather more modest improvements in strength and modulus were reported for glassy m-phenylenediamine DGEBA nanocomposites [58].

In general, modulus is the primary mechanical property that is improved through the inclusion of exfoliated silicates. The degree of improvement can be ascribed to the high aspect ratio of the exfoliated platelets. It is as-sumed [12, 55] that the reinforcement provided through exfoliation is due to shear deformation and stress transfer to the platelet particles. Zilg et al. [12, 74] have characterized the modulus and tensile strength of various hex-ahydrophthalic anhydride (HHPA) cured DGEBA nanocomposites based on different smectites and different layered silicate modifications. All systems investigated exhibited an increase in Young's modulus, although in several cases, decreases in tensile strength and failure strain were observed. It is thought that the loss in tensile strength might be related to an inhomoge-neous network density due to different cure rates of the intergallery and extragallery reaction. This may lead to internal stresses in the material, which reduces the resistance to mechanical strain. Our work [40, 41, 110] on glassy high performance nanocomposites based upon diethylene diamine cured epoxy resins of different structures and functionalities reported improved toughness and stiffness for each resin system, which could be further im-

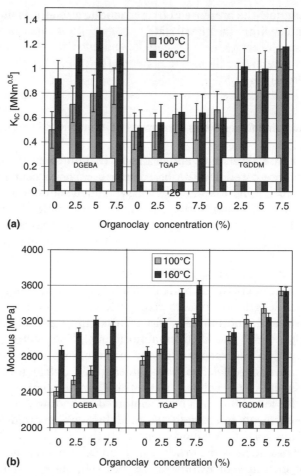

Fig. 20a,b Fracture toughness (**a**) and Young's modulus (**b**) of various diethylene diamine cured high performance epoxy nanocomposites cured at different temperatures

proved in some cases through better exfoliation of the layered silicate at higher cure temperatures as shown in Fig. 20a and Fig. 20b. Similar results were found by Kornmann et al. [44], along with a slightly decreased tensile strength and elongation at break for a series of glassy, highly crosslinked (TGDDM/DDS) nanocomposites.

Massam and Pinnavaia [72] investigated the compressive properties of intercalated and exfoliated glassy epoxy nanocomposites. In their work it was found that exfoliated systems gave significant improvements in compressive strength and modulus with increasing layered silicate concentration in the range of 0–10 wt % clay, whilst nanocomposite systems did not improve reinforcement under compression. The better improvement in the exfoliated nanocomposite was explained by the fact that each nanolayer participated fully in the reinforcement. Different degrees of improvement were reported for differing amounts of exfoliation, and also due to other factors such as different aspect ratios and charge layer densities.

Zerda and Lesser [111] also investigated the behaviour of intercalated glassy epoxy nanocomposites under compression. Their work also showed that compressive strength and modulus of the intercalated epoxy nanocomposite systems did not change noticeably. However, the yielding mechanism was found to be different in the nanocomposite compared to the unfilled epoxy. Whilst the unfilled system exhibited a gross yield behaviour with no apparent void formation, the nanocomposite yielded in shear, as evidenced by scattering of visible light by voids in the layered silicate aggregates.

7.3.2
Fracture Properties

Whilst many papers focus on the improvement of flexural properties of nanocomposites, less work has been reported regarding the fracture behaviour of these materials. However, the work presented to date has shown that many layered silicate nanocomposites show simultaneous improvement in both fracture toughness and stiffness, although elsewhere in materials science it is often found that improvement in one property generally occurs at the expense of the other. The study by Zilg et al. [12] showed that a well-dispersed *intercalated* epoxy nanocomposite primarily improved the toughness, whereas the well *exfoliated* material largely contributed to increasing the stiffness of the material. It is well known [111] that toughening occurs within a specific size range of the reinforcement. Whilst the fully-dispersed, nanometer-dimensions of the layers is unlikely to provide a toughening mechanism, the lateral micron-length silicate tactoids may provide such toughening through a crack bridging mechanism and increased fracture surface area.

Improved fracture toughness has also been reported for other intercalated or partially exfoliated epoxy nanocomposite systems [44, 74]. Zerda and Lesser [111, 112] have investigated the fracture behaviour of intercalated DGEBA/poly(oxypropylene) diamines (Jeffamine D-230, $x = 5$–6) nanocomposites. The material investigated showed a modest increase in modulus, alongside a significant decrease in ultimate stress and strain at failure. The fracture behaviour of these materials represented by the stress intensity factor, K_{IC}, showed significant improvements for layered silicate concentrations of 3.5 vol % and above, from initial values of 0.9 MPa/m^2 to 1.5 MPa/m^2 (3.5 vol % organo-clay). This increase in fracture toughness was ascribed to a decrease in the tactoid, inter-particle distance. SEM images of the fracture surfaces showed a more tortuous path of crack propagation around areas of high silicate concentration in the nanocomposite compared to the neat system. The creation of additional surface areas on crack propagation was thus

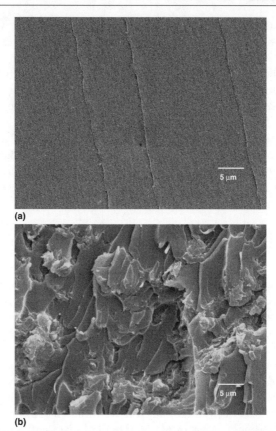

Fig. 21a,b Comparison of fracture surfaces of: (**a**) neat diethyltoluene diamine cured tetraglycedyl diamino diphenylmethane resin system (top) and (**b**) its corresponding nanocomposite containing 7.5% octadecyl ammonium modified layered silicate

assumed to be the primary factor for the toughening effect. Similar findings were reported for nanocomposites based upon unsaturated polyester [113] or high functionality epoxy resin nanocomposites [114]. A comparison of fracture surfaces of both neat and highly intercalated layered silicate containing epoxy systems is shown in Fig. 21a and b.

High strain rate impact strength has been less studied, but has also been shown to be increased by incorporation of layered silicates. Basra et al. [115] performed Charpy impact tests on treated clays that showed an increase in d-spacing (intercalation) of some 20 Å. It was found that there was a maximum in impact strength, with the peak value occurring at about 0.5 vol % treated clay, whilst the impact strength increased by some 137%, subsequently decreasing in value until 9 vol % clay, when it is again the same as the original resin. The decrease above 0.5 vol % was ascribed to clay agglomeration and untreated clays which intercalate much less, are found to produce concomitantly less improvement. In both cases, the nanocomposites ductility or strain-to-break is decreased. Similar maxima in impact strength at around 1 vol % have been reported elsewhere [116].

7.4
Dimensional Stability

Massam et al. [72] investigated the thermal expansion coefficient, α, of a series of polyoxyalkylene amine cured DGEBA layered silicate nanocomposites. Measurements in the range of 40–120 °C showed reduced expansion coefficients for both the rubbery and the glassy state. A decrease in the expansion coefficient by 27% was reported for the 5 vol % nanocomposite in the glassy state. A monotonically decreasing expansion coefficient with increasing layered silicate concentration was found in the rubbery state, with an organoclay loading of 15 vol % showing a 20% reduction in α.

7.5
Water Uptake and Solvent Resistance

A comprehensive study by Massam et al. [31, 72] investigated the resistance of glassy DGEBA based nanocomposites towards organic solvents and water. The absorption of methanol, ethanol and propanol was faster in the neat epoxy system, compared with the nanocomposite. Furthermore, the mechanical properties of the neat resin systems were more affected by the absorbed solvent. For example, after 30 days of exposure to methanol, a neat epoxy system became rubbery, whilst the related composite material appeared unaffected. A pristine polymer submerged in propanol absorbed more than 2.5 times than the nanocomposite, and began to crack and break up, whilst the shape and texture of the nanocomposite remained unchanged. In water, however, the rate of absorption was reduced, with little change in equilibrium uptake. The absorption mechanism or the role of polarity of the solvent was not discussed.

Gensler et al. [117] reported significantly reduced water vapor permeability for a hexahydrophtalic anhydride cured DGEBA nanocomposite. The nanofiller used in this work was an organically-modified hydrotalcite which, in contrast to layered silicates, has a positive layer charge in the gallery which is counterbalanced by anions. The water vapor permeability of the highly intercalated nanocomposite was five to ten times reduced at a content of 3 wt % and 5 wt % hydrotalcites, respectively, when compared with the neat polymer.

Recent work on highly intercalated and ordered exfoliated glassy epoxy nanocomposites [118] found that the neat epoxy systems generally absorb more water than the nanocomposites. A monotonic decrease in water sorption with increasing clay concentration, however, was not observed. In contrast to the work by Massam et al. [31, 72], the rate of absorption remained relatively unaffected by organoclay addition.

Shah et al. [105] recently published a study on moisture uptake of vinyl ester-based layered silicate nanocomposites. Although the moisture diffusivity decreased with the addition of organoclay, the equilibrium moisture uptake was found to increase or remain unchanged by the amount of added layered silicate. The diffusion coefficient was reduced from 0.022×10^6 mm^2/s to 0.015×10^6 mm^2/s with the addition of 5 wt % of vinyl monomer clay and Cloisite 10A® (natural montmorillonite modified with a quaternary ammonium salt, benzyltallowdimethylammonium), whilst the equilibrium water uptake increased from 0.012 wt % for the neat material to 0.021 wt % with clay. The increased equilibrium water uptake was explained by the hydrophilic behaviour of the clay which persists, even though the surface has been treated. Higher concentrations of layered silicate may lead to aggregates or tactoids of layered silicate with less exposed surface area, leading to a negative deviation from the linear relationship between equilibrium water absorbed and organoclay concentration. Furthermore, it was found that the diffusion coefficient did not differ significantly between two different clay modifications, which showed different degrees of separation of the layered silicates. The decreased diffusivity was thus ascribed to the restricted motion of polymer chains that are tethered to the clay particles.

7.6
Thermal Stability and Flammability

Thermogravimetric analysis (TGA) is the most commonly-used method to investigate the thermal stability of polymers, which is also an important property for the flammability performance of a material [119–121]. To date, the thermal stability of epoxy nanocomposites has been mostly investigated for DGEBA-based systems with the onset and peak degradation temperature of TGA traces and the char level being the main parameters reported. Lee and Jang [122] found improved thermal stability for intercalated epoxy nanocomposites synthesized by emulsion polymerization of unmodified layered silicate, as indicated by a shift in the onset of thermal decomposition (in a nitrogen atmosphere) towards higher temperatures. Wang and Pinnavaia [37] compared TGA measurements (also under a nitrogen atmosphere) of intercalated and exfoliated organically modified magadiite nanocomposites. Whilst the intercalated epoxy nanocomposite showed a low temperature weight loss at about 200 °C, indicative of the thermal decomposition of the clay modifier, the exfoliated nanocomposite did not show such a low onset temperature for weight loss and it was proposed that the interlayer exchanged ions were incorporated into the polymer network. Recent work by Gu and Liang [123] investigated the thermal degradation of DGEBA-based nanocom-

posites containing 2 and 10 wt % octadecylammonium modified montmorillonite in air and nitrogen environments, respectively. It was found that the
10% nanocomposite had the lowest degradation temperature, whereas values
for the nanocomposites containing 2 wt % organo-clay were higher than the
neat resin.

The incorporation of organically modified layered silicates into high performance thermosetting systems with high T_gs and good thermal stability may decrease the thermal stability of the overall material in terms of
initial degradation temperature. Xie et al. [124] recently reported a detailed investigation of the non-oxidative thermal degradation chemistry of
quaternary alkylammonium-modified montmorillonite. The onset of true
organic decomposition (rather than water desorption which could be observed at lower temperatures) was found to be 180 °C, and the decomposition process was divided into four stages: the desorption of water and
other low molecular weight species (below 180 °C) the decomposition of organic substances (200–500 °C), the dehydroxylation of the aluminosilicate
(500–700 °C) and residual organic carbonaceous evolution at 700–1000 °C.
Furthermore, the work suggested a Hoffmann elimination reaction as the
mechanism of the initial thermal degradation. Recently reported TGA measurements [118] on high performance (octadecyl ammonium modification
based) epoxy nanocomposites with T_gs of 175 °C in an inert nitrogen atmosphere found that the nanocomposites showed a slightly reduced thermal stability compared to the neat epoxy systems, as indicated by a decreased degradation onset temperature in the order of 5–10 °C. In addition,
cone calorimetric measurements of these systems [125] showed synergistic improvement in fire retardancy, as indicated by a reduced peak release
rate.

The mechanism of the improvement of thermal stability in polymer
nanocomposites is not fully understood. It is often stated [126–129] that
enhanced thermal stability is due to improved barrier properties and the
torturous path for volatile decomposition products, which hinders their diffusion to the surface material where they are combusted. Other mechanisms
have been proposed, for example, Zhu et al. [130] recently proposed that
for polypropylene-clay nanocomposites, it was the structural iron in the dispersed clay that improved thermal stability by acting as a trap for radicals at
high temperatures.

The flammability of nanocomposites has been the subject of various studies, largely by Gilman and coworkers [121, 129, 131, 132]. Flammability properties are most often investigated using a cone calorimeter, a method where
properties relevant to combustion such as heat release rate (HRR), and carbon monoxide yield during burning of a material are measured. The group
of Gilman et al. [121, 132] presented results of flammability studies of a number of thermosetting systems including cyanate esters [132] and vinyl esters
and epoxies [121, 133]. Table 2 shows the results of combustion of different

Table 2 Cone calorimeter data for modified bisphenol A vinyl ester (Mod-Bis-A Vinyl Ester), bisphenol A novolac vinyl ester (Bis-/Novolac Vinyl Ester) and methylenedianiline and benzyldimaine (BDMA) cured epoxy resins and their intercalated nanocomposites (*) containing 6% dimethyl dioctadecylammonium-exchanged montmorillonite. Heat flux = 35 kW/m², HRR = heat release rate, MLR = mass loss rate, H_c = heat of combustion, SEA = specific extinction area [121]

Sample	Residue Yield []%	Peak HRR [kW/m²] (Δ%)	Mean HRR [kW/m²] (Δ%)	Mean MLR [g/s m²] (Δ%)	Mean H_c [MJ/kg]	Mean SEA [m²/kg]	Mean CO yield [kg/kg]
Mod-Bis-A Vinyl Ester	0	879	598	26	23	1360	0.06
Mod-Bis-A Vinyl Ester*	8	656 (25%)	365 (39%)	18 (30%)	20	1300	0.06
Bis-/Novolac Vinyl Ester	2	977	628	29	21	1380	0.06
Bis-/Novolac Vinyl Ester*	9	596 (39%)	352 (44%)	18 (39%)	20	1400	0.06
DGEBA/DDM	11	1296	767	36	26	1340	0.07
DGEBA/DDM*	19	773 (40%)	540 (29%)	24 (33%)	26	1480	0.06
DGEBA /BDMA	3	1336	775	34	28	1260	0.06
DGEBA /BDMA*	10 (42%)	769 (35%)	509 (38%)	21	30	1330	0.06

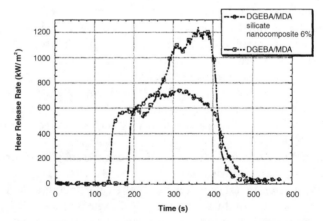

Fig. 22 Heat release rate data for DGEBA epoxy resin cured by methylenedianiline (MDA with and without nanocomposite (6 wt % clay). The clay was a montmorillonite treated with dimethyl ditallow ammonium ions. The cone calorimeter was run at a heat flux of 35 kW/m² [133]

thermosets obtained from cone calorimetry [121] and a typical heat release rate curve for a neat epoxy and a nanocomposite, as is shown in Fig. 22. It can be seen that the peak and average release rate, as well as mass loss rates, are all significantly decreased upon organoclay addition. Furthermore, no increase in heat of combustion, specific extinction area (soot) or CO yields was seen. A slightly shorter time to ignition occurs because of the instability of the quaternary ammonium organo-ion, as mentioned above. The mechanism of improved flame retardancy is not yet fully understood and there is no general agreement about which structure (intercalated or exfoliated), gives the best flammability properties [121]. It was found that reduced mass loss rate occurred only when the sample surface was partially covered with char. It is believed that the nanocomposite structure in the char acts as an insulator for both heat and mass transfer. TEM images of the char of different nanocomposite systems (thermoplastics and thermosets) showed that the interlayer spacing of the char was constant (1.3 nm), independent of the chemical structure of the nanocomposite. The nanocomposite strategy for flame retardation offers a number of benefits, such as improved flammability along with improved mechanical properties, whilst being more environmentally-friendly compared to other common flame retardants for relatively low concentrations and costs. It is believed that the additional use of layered silicates for improved flammability performance may lead to the removal of significant portions of conventionally-used flame retardants [129], although it is likely that layered silicates on their own are not sufficient for this purpose.

7.7
Optical Properties

Layered silicate nanocomposites are often found to exhibit good transparency. Wang et al. [54] compared the optical properties of organically modified magadiite and smectite based epoxy nanocomposites of a 1 mm thick sample with a concentration of 10 wt % layered silicate. Both systems showed good optical properties. A better transparency of the magadiite nanocomposite, however, was related to either better exfoliation or a better match with the refractive index of the organic matrix. Comparison of intercalated and exfoliated epoxy nanocomposites of up to 20 wt % organoclay concentration by Brown et al. [61] showed good transparency for all exfoliated systems, as well as for low concentrations of intercalated layered silicates. Recent work by Salahuddin et al. [49] showed that films of highly filled epoxy nanocomposites (up to 70 wt % layered silicate) also show good clarity due to the molecular level of the dispersion.

8
Ternary Layered Silicate Nanocomposite Systems

As the field of thermosetting layered silicate nanocomposites is still relatively new, the major work to date has focused on the understanding of
morphology, processing conditions and properties of less complex binary
nanocomposite systems. However, the promising results reported for this new
class of material have recently encouraged research in nanocomposites where
it is a supplementary additive, used in combination with other phases such as
fibres, rubbers or hyperbranched polymers.

8.1
Epoxy fiber nanocomposites

Recent work [134, 135] on DGEBA/diethyltoluenediamine based fibre nanocomposites has shown improvement in the mode I fracture toughness
through the incorporation of highly intercalated octadecylammonium modified layered silicates. The panels were synthesized by painting the nanocomposite premix (resin/hardener/organo-clay blend) onto the unidirectional
fibre cloth. The prepregs were then aligned in a mould and stacked to
a 28-ply laminate. A piece of 13 μm thick polyimide foil was placed in the
mid-thickness of the fibre plies as a crack initiator for the mode I fracture
toughness test, the panel then cured in a hot press. Figure 23 shows improvement in maximum load and fracture energy (G_{IC}) as a function of
organoclay concentration. Timmermann et al. [136] reported improved resis-

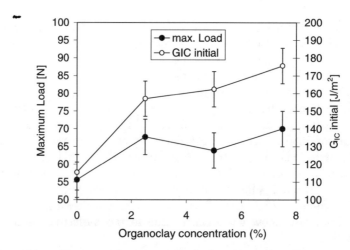

Fig. 23 In-plane resistance and maximum load of DGEBA carbon fibre nanocomposite as
determined from mode I fracture toughness tests [135]

tance in thermal cycling induced cracking of epoxy/fibre composites through incorporation of exfoliated layered silicates.

The work by Rice et al. [137] investigated the matrix-dominant properties of a bisphenol F/epichlorhydrin epoxy resin layered silicate fibre composites through four-point flexure measurements, and found no significant increase in z-axis properties. Little improvement was found for a fibre composite with low organoclay concentration. Composite systems of higher organoclay concentrations even showed a decreased flexural strength, which was ascribed to an increased void content in the matrix. In their work, the composite material was produced using a wet winding (WW) method and resin film infusion (RFI) technique, and concluded that the WW method showed more promise than RFI. This was explained by a filtering effect of the clay particles, and thus the filler was unevenly distributed throughout the polymer matrix. Consequently, the optimal processing of ternary carbon fibre nanocomposites is critical to achieving significant improvement in the composite materials. Understanding the resin flow, kinetics and gel times are also key to optimizing the cure profiles of the thermosetting systems. The effect of the carbon fiber upon the dispersion and exfoliation of the layered silicate is also significant and apparently determined by the method of fabrication.

8.2
Ternary systems consisting of a layered silicate, epoxy and a third polymeric component

Most glassy thermosetting materials are themselves intrinsically brittle and some form of toughening is often required. This includes incorporation of materials such as liquid rubbers which phase separate and improve toughness by a range of mechanisms such as increased yielding, cavitation and crack blunting. A downside of this toughening method is decreased modulus of the thermoset material due to the rubbery particles, a problem compounded by the often-incomplete phase separation of the rubber which results in a lower glass transition temperature of the rubber-plasticised matrix that remains. Recent work [138, 139] has investigated a new strategy to produce toughened epoxy resins which maintain a high modulus through ternary blends of a rubbery phase, layered silicate and rigid epoxy. Both a liquid reactive rubber and a hyperbranched epoxy resin have been considered as the toughening phase. The use of hyperbranched epoxy resins, rather than classical rubber materials as the toughening agent, is relatively new and has shown to effectively toughen glassy epoxy matrices [17, 18].

Toughened epoxy resins and epoxy nanocomposite systems were synthesized using DGEBA resin, diethylene diamine hardener, octadecylammonium modified montmorillonite and epoxy functional dendritic hyperbranched polymer (Boltorn E1, Perstorp Speciality Chemicals, Sweden) with an epoxy equivalent weight of $\sim 875\,g/eq$ and a molecular weight of $\sim 10\,500\,g/mol$.

Table 3 Properties of ternary nanocomposite comprising DGEBA epoxy resin, hyperbranched epoxy resin and octadecylammonium-modified organo-silicate

Blend composition Epoxy/HBP/clay	Estimated d-spacing [Å]	T_g [°C]	Flexural strength [MPa]	Flexural Modulus [MPa]	Impact strength [J/m]
100/0/0	–	192	112	2920	740
100/15/0	–	182	109	2510	2250
100/0/5	90–100	206	146	4090	1060
100/15/5	90–100	192	135	3630	1540
0/100/5	120	– 10	–	–	–

Becker, Simon

This hyperbranched polymer (HBP) consists of a highly branched aliphatic polyester backbone with an average of 11 reactive epoxy groups per molecule. Investigation of the morphologies of these materials showed that the DGEBA nanocomposite had a well-dispersed structure with tactoids of layered silicates still remaining, with an average d-spacing of 90–100 Å. The microstructure of ternary epoxy/HBP/layered silicate systems consisted of distinct regions of highly intercalated, layered silicates along with spheres of HBP of approximately 0.8–1 μm. It was found that the presence of the clay has little effect on HBP phase separation. Table 3 summarizes the average d-spacing, glass transition temperatures and mechanical properties of the systems investigated. Although both the clay and the HBP show a toughening effect on the epoxy matrix, in the ternary blend the overall toughness is less than the toughness of the HBP/epoxy system alone.

Lelarge et al. [138] investigated the ternary system involving DGEBA with a CTBN-rubbers of differing polarity and a montmorillonite surface-treated with the octadecyl ammonium organo-ion. The system was investigated in some detail, from the catalytic effect both the treated-clay and the rubber had on the epoxy reaction, to the effects of the rubber on the degree of intercalation and the influence of the layered silicate of phase separation, gelation and vitrification of the epoxy network. It was found that the clay and epoxy formed good, highly intercalated structures (d-spacings greater than 70 Å). However, this was reduced in the final ternary systems because not all rubber phase separates (and this was investigated as a function of rubber polarity) in such a system and some remains soluble in the crosslinked matrix. Although the rubber can intercalate the layered silicate alone, it does not do so to the same degree as the epoxy resin. Indeed, in the ternary blend, the clay remained in the epoxy-rich phase, as clearly seen in Fig. 24a and 24b, whilst the rubber phase separated into fine particles, as in the binary epoxy – rubber

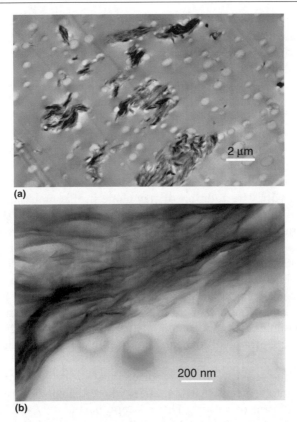

Fig. 24a,b TEM micrographs of ternary epoxy, CTBN rubber and octadecyl ammonium-treated montmorillonite nanocomposites which were investigated with different polarity rubbers. (a) shows a broad view of the less polar rubber and the clay tactoids, clearly separated and (b) shows a close up of the more polar rubber close to silicate layers

mixture. In-situ studies of various properties found that the clay essentially delaminated to the final degree possible, prior to phase separation, which itself occurs before gelation. Flexural testing indicated that the clay was able to retrieve some of the modulus lost by rubber addition. Although toughness of the resin increased with the addition of clay alone and, of course, rubber alone – the effect was not synergistic, the final toughness laying between that of clay-alone and the higher, rubber-alone value.

Frölich et al. [140] investigated a system in which DGEBA was mixed with hydroxy-terminated poly(propylene oxide-block-ethylene oxide) as the rubber, with the nanoclay being a synthetic fluorohectorite treated with bis (2-hydroxyethyl) methyl tallow alkylammonium ions. The clay was first blended with rubber, before being dispersed into the reactive epoxy mixture. Modification of the rubber allowed variation in miscibility and differing morphologies and properties. If the rubber was miscible, the intercalated clay led to improved toughness. If the rubber is sufficiently modified, such as with methyl stearate, then distinct phase-separated particles and silicate layers co-reside. The phase-separated morphology leads to a significant increase in toughness, with only a modest modulus decrease.

Isik et al. [116] investigated the use of polyether polyol to modify DGEBA/triethylenetetramine/organo-treated montmorillonite nanocomposites where the layered silicate was modified with bis (2-hydroxyethyl) methyl tallow alkyl ammonium ion. The polyether polyol was used in concentrations of up to 7 wt % and formed domains in the epoxy of some 0.6 μm (1 wt % polyol) to 1.6 μm (5 wt % polyol). The addition of both clay and polyol alone in the epoxy are found to increase toughness due to cavitation (250% increase for 7 wt % polyol). The addition of clay alone in this material, which intercalates to tactoids which show a 20 Å increase in d-spacing, also results in an increase in impact strength of some 170% for low contents (ca. 1 wt %), but a decrease at higher contents. The combination of both polyol and clay, as with the other ternary systems reported to date, does not show a further synergistic increase. The addition of clay in the ternary system causing a decrease in impact strength, although it is still generally greater than the neat epoxy system. It is found that the polyol itself does not enter the clay galleries. The modulus is similar or slightly greater than that of the neat resins, although the highest polyol concentrations (even at high clay concentrations) cause reduced values.

An interesting recent variation reported involves the inclusion of clay with epoxy resins into a thermoplastic matrix, where the epoxy resin now serves as a reactive diluent or solvent. This concept, particularly with epoxies, was pioneered some years ago [141–143] for thermoplastic polymers that are difficult to process. In these cases minority additions of epoxies can be included and with appropriate miscibility, aid processing. These same systems lead to a higher composite modulus when the inclusions cure (as opposed to reduced rigidity for additions of non-reactive plasticisers) [144]. In the work involving the ternary system in question, poly(methyl methacrylate) (PMMA) is used as the thermoplastic matrix, and rather than a simple DGEBA epoxy being added, a mixture of aromatic and aliphatic epoxies were used to ensure a low glass transition epoxy matrix. By creating an epoxy phase with a T_g of 69 °C, lower than that of PMMA, rubber toughening is thus also possible. The clay was once again the bis (2-hydroxyethyl) methyl tallow alkylammonium ion-treated montmorillonite. The various binary blends were also investigated, as well as the ternary one. Clay was found to become intercalated in both the PMMA/clay and epoxy/clay binary systems, with a decrease in impact strength and strain-to-failure in the PMMA-clay material. The PMMA/epoxy binary blend itself lead to toughening, the 10–30 wt % cured epoxy spheres ranging in diameter from 0.1–0.6 μms in size (although the viscosities of the components meant similar-sized voids were also seen). Nonetheless, a modest increase of some 6% impact strength was observed. In the ternary system, the layered silicates were pre-intercalated into the reactive epoxy solvent, before inclusion with the PMMA and they remained encased in the epoxy phase during cure. They resulted in larger epoxy domains of ca. 1–10 μm (compared to less than 1 μm for epoxy in PMMA alone). The degree of dispersion was influenced by the ratio of epoxy-to-clay and this makes direct determination of the effect of including clay in the epoxy matrix difficult, as much of the failure occurs around the epoxy-PMMA interface. However, modulus was still found to increase due to the presence of the clay in the epoxy phase. This concept of including the nanoclay in a reactive diluent has much promise in other thermoplastic-thermoset systems, particularly if the miscibility of the reactive solvent can be manipulated to aid dispersion of the clay more completely throughout the thermoplastic matrix.

Thermoset blends need not always be epoxy and thermoplastic or rubber. Over the years, interpenetrating polymer networks (IPNs) of various types have become widely studied, where mixtures of reactive materials (where at least one monomer is polyfunctional) can be simultaneously or sequentially polymerized. In these types of systems, varying degrees of miscibility may be trapped by reactive chains and permanent entanglements, and many different properties can occur. The area has been well reviewed, one recent example being [145]. There have been few reports of nanoclays introduced into such systems. One recent paper in this vein is the work of Karger-Kocsis et al. [146] who had previously reported work with a vinyl ester resin and epoxy IPN which were found to be tough but were of low modulus and glass transition [147]. It was thought that inclusion of a nanocomposite phase could contribute favourably to both modulus and glass transition, as described above as the motivation for including layered silicates into rubber-toughened epoxies. Using the bis (2-hydroxyethyl) methyl tallow alkylammonium ion on montmorillonite and another synthetic clay, nanocomposites were made by adding the treated-clays to the pre-mixed equally-proportioned vinyl/ester epoxy resin mixture. However, the addition of the clays was found to further decrease thermal and fracture properties. The epoxy resin appeared to encapsulate the clays – possibly in a non-stoichiometric manner, which showed some level of intercalation. A small amount of clay (some 5 wt %) was found to increase fracture toughness but this was not a nano-phenomena, but rather due to the toughening effects of the softer interphase between encapsulated clay and the matrix.

9
Conclusions and Future Directions

Epoxy – layered silicate composites have attracted much interest and extensive research within the area of nanocomposites. They are relatively easy to make, and since they comprise a reactive system such as epoxies (compared to melt blends) they are also applicable in a range of areas – from adhesives to coatings, from microelectronic applications to composite systems. The addition of nanoclay has the potential to have a range of benefits from increased modulus, strength, fracture toughness, impact resistance, gas and liquid barrier, flame retardance and so on, all at moderately low concentrations of about 3–5 wt %. The ability also to improve toughness, particularly in highly crosslinked epoxies, was perhaps not totally expected based on the loss of ductility seen in thermoplastic matrices such as fully-exfoliated clays observed in nylon 6 matrices. The low concentration means that there are few negative implications for processing. The fact that most reported materials do not show full exfoliation, yet often exhibit much improved properties, indicates exfoliation is not required to obtain many of the desired properties. In many cases, the degree of interlayer spacing in epoxy nanocomposites increases beyond the 60–80 Å which is the lowest scattering angles possible in most laboratories WAXD devices. TEM (and the much less reported) SAXD become important, where layered structures of some 130–140 Å can still be seen.

   Research in this area continues to be very active. Much of the current work involves seeking a fundamental understanding of the manner in which materials intercalate and exfoliate, and the factors which allow this to be best achieved. This is related, of course, to the effect the various nanoclay treatments on reaction chemistry, as well as the final physical properties. In particular, the influence on properties such as the glass transition and how the degree of dispersion and attachment influence this remains of much interest. In this regard, much of the work to date has involved a limited range of resins (DGEBA) and surface treatments (alkylammonium or hydroxy-ammonium ions). It is not yet clear as to the extent which further chemical design and synthesis of organo-ions can progress properties. As well as attachment to the epoxy matrix, the use of non-ammonium organo-ions may have some advantages in high temperature stability [148]. The use of other non-organoion based additives, such as copolymers in untreated clays, has not been widely examined at this stage.

   Fracture toughness, always important in composites and adhesives, remains an important issue, as does a better understanding of the relative importance of intercalation/exfoliation and the influence of their size-scale (micron vs. nanometer). Indeed, the effect of the size-scale of materials from individual layers to tactoids to clusters to larger clay clumps, many of which seem to exist simultaneously in a sample, is still not fully understood or able to be independently manipulated. The ability to make full use of the anisotropy of the clays in thermoset components by appropriate processing techniques also is of interest. Likewise, the ability to usefully add greater concentrations of clay in certain applications for particular property balances (as opposed to widely-used 3–5 wt %) may also allow a greater range of properties.

There remains good work to be done in characterization. The standard wide angle X-ray diffraction and TEM techniques remain the necessary, basic tools of characterization. Further use could be made of various ablative techniques such as plasmas to reveal structures. Even in these there remains more sophisticated analysis possible, such as detailed in other nanocomposite systems with WAXD [149] and TEM [150]. The use of other important techniques such as small angle X-ray scattering is becoming increasingly used, particularly important in epoxy nanocomposites where very good intercalation is often observed which is outside the window of traditional WAXD measurement. It is probably fair to say that the use of various types and modes of atomic force microscopy has also not been utilised to their full extent at this time, which may also be aided by appropriate surface treatments (to etch and expose surface morphology). The relative ease of AFM compared to the specialist skills required, and time necessary, for microtoming and TEM, may be important. Other properties, such as the effect of clay addition on friction, thermal conductivity and free volume and the like have not been widely examined but may be useful in certain applications. Much of the advantage of nanoclays is the multiplicity of effects the clay can have on chemistry, mechanical properties, fire performance, barrier properties and so on, and there is clearly an interplay between them and need to understand the effects on a small-scale.

Ternary systems are becoming more widely reported with, in addition to epoxy and clay, other materials being present such as rubber, thermoplastic or fibres. Synergies need to be sought. Likewise, the addition of additives such as flame retardants, either physically blended, or covalently-incorporated with the epoxy or amine need to be examined in nanocomposites, since this is one of the most important, ongoing requirements of transport industries such as aerospace.

Acknowledgements Ole Becker acknowledges financial support received from the Monash University Postgraduate Publications Award, as well as for a postgraduate scholarship. GPS acknowledges support from the Australian Research Council for some of this work.

References

1. Pinnavaia JT (1983) Science 220:365
2. Mardis WS (1984) JAOCS 61:382
3. Kemnetz SJ, Still AL, Cody CA, Schwindt R (1989) J Coat Technol 61:47
4. Zanetti M, Lomakin S, Camino G (2000) Macromol Mater Eng 279:1
5. Solomon DH, Hawthorne DG (1983) Chemistry of Pigments and Fillers. John Wiley & Sons, New York

6. Hawley GC (1987) Handbook of Reinforcements for Plastics, ed. JV Milewski, Katz, HS,: Van Nostrand Reinhold Company, New York
7. Blumstein A (1961) Bull Soc Chim Fr 5:899
8. Okada A, Kawasumi M, Usuki A, Kojima Y, Kurauchi T, Kamigaito K (1990) Mat Res Symp Proc 171:45
9. Usuki A, Kojima Y, Kawasumi M, Okada A, Fukushima Y, Kurauchi T, Kamigaito O (1993) J Mater Res 8:1179
10. Usuki A, Kojima Y, Kawasumi M, Okada A, Fukushima Y, Kurauchi T, Kamigaito O (1993) J Mater Res 8:1185
11. Yano K, Usuki A, Okada A, Kurauchi T, Kamigaito O (1993) J Polymer Science: Pt A: Polym Chem 31:2493
12. Zilg C, Mülhaupt R, Finter J (1999) Macromol Chem Phys 200:661
13. Young RJ (1986) Rigid-particulate reinforced thermosetting polymers in Structural Adhesives – Developments in Resins and Primers. Kinloch AJ (ed) Elsevier Applied Science Publishers: London
14. Spanoudakis J, Young RJ (1984) J Mat Sci 19:473
15. Ellis B (1993) Chemistry and Technology of Epoxy Resins, Blackie Academic, London
16. Sankaran S (1990) J Appl Polym Sci 39:1635
17. Ratna D, Varley R, Singh Raman RK, Simon GP (2003) J Mat Sci 38:147
18. Boogh L, Pettersson B, Manson J-A (1999) Polymer 40:2249
19. Bucknall CB, Partridge IK (1983) Polymer 24:639
20. Varley R, Thermoplastic Modification of a Trifunctional Epoxy Resin System, in PhD thesis, Department of Materials Engineering. 1998, Monash University: Clayton, Melbourne
21. Kim J-K, Mai Y-W (1998) Engineered Interfaces in Fiber Reinforced Composites. Elsevier, Amsterdam
22. Ivens J, Debaere P, McGoldrick C, Verpoest I, Van Der Vleuten P (1994) Composites 25:139
23. Chung WC, Jang BC, Chang TC, Hwang LR, Wilcox RC (1989) Mat Sci Eng A112:157
24. Zhong W, Jang BZ (1998) Key Eng Mat 141–143:169
25. Garcia R, Evans RE, Palmer RJ (1987) Structural Property Improvements Through Hybridized Composites. Toughened Composites. Johnston NJ (ed) Philadelphia: American Society for Testing Materials. 397.
26. Yamashita S, Hatta H, Takei T, Sugano T (1992) J Composite Mater 26:1254
27. Jang BZ (1991) Sci and Eng of Composite Mater 2:29
28. Sohn MS, Hu XZ (1998) Key Eng Mater 145–149:727
29. Hoffmann U, Endell K, Wilm D (1933) Z Krist. 86:340
30. Alexandre M, Dubois P (2000) Mater Sci Eng 28:1
31. Wang Z, Massam J, Pinnavaia TJ (2000) Epoxy-Clay Nanocomposites, in Polymer-Clay Nanocomposites. Pinnavaia TJ, Beall GW (eds), Wiley, Chichester. p. 127
32. Theng BKG (1974) The Chemistry of Clay-Organic Reactions. Wiley, New York
33. Bailey SW, Brindley GW, Johns WD, Marti RT, Ross M (1971) Clays and Clay Minerals 19:129
34. Theng BKG (1979) Formation and Properties of Clay-Polymer Complexes. Development in Soil Science 9, Elsevier, New York
35. Fletcher RA, Bibby DM (1987) Clays and Clay Minerals 35:318
36. Giannelis EP (1992) Materials & Design 13:100
37. Wang Z, Pinnavaia JT (1998) Chem Mater 10:1820
38. Akelah A, Kelly P, Qutubuddin S, Moet A (1994) Clay Minerals 29:169
39. Morgan AB, Gilman JW, Jackson CL (2001) Macromolecules 34:2735

40. Becker O, Varley RJ, Simon GP (2002) Polymer 43:4365
41. Becker O, Cheng Y-B, Varley RJ, Simon GP (2003) Macromolecules 36:1616
42. Vaia RA (2000) Structural Characterization of Polymer-Layered Silicate Nanocomposites, in Polymer-clay nanocomposites, Pinnavaia JT, Beall GW (eds), John Wiley & Sons, p 229
43. Morgan AB, Gilman JW, Jackson CL (2000) Characterization of Polymer-Clay Nanocomposites: XRD vs. TEM in 219th ACS National Meeting. San Francisco, California
44. Kornmann X, Thomann R, Mülhaupt R, Finter J, Berglund LA, (2002) Polym Eng Sci 42:1815
45. Chin I-J, Thurn-Albrecht T, Kim C-H, Russell TP, Wang J (2001) Polymer 42:5947
46. Tolle TB, Anderson DP, (2002) Composites Science and Technology 62:1033
47. Chen C, Curliss D (2003) Nanotechnology 14:643
48. Kornmann X, Lindberg H, Berglund LA (2001) Polymer 42:1303
49. Salahuddin N, Moet A, Hiltner A, Baer E (2002) Eur Polym J 38:1477
50. Reichert P, Nitz H, Klinke S, Brandsch R, Thomann R, Mülhaupt R (2000) Macrom Mater Eng 275:8
51. VanderHart DL, Asano A, Gilman JW (2001) Macromolecules 34:3819
52. Devaux E, Bourbigot S, Achari AE (2002) J Appl Polym Sci 86:2416
53. Kojima Y, Usuki A, Kawasumi M, Okada A, Kurauchi T, Kamigaito (1993) J Polym Sci Pt A: Polym Chem 31:983
54. Wang Z, Lan T, Pinnavaia JT (1996) Chem Mater 8:2200
55. Lan T, Pinnavaia TJ (1994) Chem Mater 6:2216
56. Yasmin A, Abot JL, Daniel IM (2003) Scripta Materiala 49:81
57. Butzloff P, D'Souza NA (2003) Epoxy + Montmorillonite Nanocomposites: Effects of Water, Ultrasound, and Stoichiometry on Aggregates. Proc ANTEC
58. Lan T, Kaviratna PD, Pinnavaia TJ (1995) Chem Mater 7:2144
59. Becker O, Bourdonnay R, Halley PJ, Simon GP (2003) Polym Eng Sci
60. Jiankun L, Yucai K, Zongneng Q, Xiao-Su Y, (2001) J Polym Sci Part B: Polym Phys 39:115
61. Brown JM, Curliss D, Vaia RA (2000) Chem Mater 12:3376
62. Chen C, Curliss D, (2003) J Appl Polym Sci 90:2276
63. Chen JS, Poliks MD, Ober CK, Zhang Y, Wiesner U, Giannelis E (2002) Polymer 43:4895
64. Kong D, Park CE, (2003) Chem Mater 15:419
65. Vaia RK, Jandt KD, Kramer EJ, Giannelis EP (1995) Macromolecules 28:8080
66. Krishnamoorti R, Giannelis EP, (1997) Macromolecules 30:4097
67. Park JH, Jana SC, (2003) Macromolecules 36:2758
68. Ke Y, Lü J, Yi X, Zhao J, Qi Z (2000) J Appl Polym Sci 78:808
69. Pinnavaia TJ, Lan T, Kaviratna PD, Wang Z, Shi H (1995) Clay-Reinforced Epoxy Nanocomposites: Synthesis, Properties and Mechanism of Formation. in Polymeric Materials Science and Engineering. Chicago, Illinois: American Chemical Society
70. Shi H, Lan T, Pinnavaia TJ, (1996) Chem Mater 8:1584
71. Lan T, Kaviratna PD, Pinnavaia TJ (1996) J Phys Chem Solids 57:1005
72. Massam J, Pinnavaia TJ (1998) Clay Nanolayer Reinforcement of a Glassy Epoxy Polymer. Mater Res Soc Symp Proc 520:232
73. Lan T, Kaviratna PD, Pinnavaia TJ (1994) Polym Mater Sci Eng 71:527
74. Zilg C, Thomann R, Finter J, Mülhaupt R (2000) Macromol Mater Eng 280/281:41
75. Messersmith PD, Giannelis EP (1994) Chem Mater 6:1719
76. Chen JJ, Cheng IJ, Chu CC (2003) Polymer Journal 35:411

77. 30B, C, Physical Properties Bulletin, Southern Clay Products, USA. http://www.nanoclay.com/data/30B.htm
78. Feng W, Ait-Kadi A, Riedl B (2003) Polym Eng Sci 42:1827
79. Kornmann X, Lindberg H, Berglund LA (2001) Polymer 42:4493
80. Chen C, Curliss D (2003) Polym Bull 49:473
81. Lan T, Wang Z, Shi H, Pinnavaia TJ (1995) Clay-epoxy nanocomposites: relationships between reinforcement properties and the extend of clay layer exfoliation. in: Polimeric Materials – Science and Engineering. Proceedings of the ACS Division of Polymeric Materials.
82. Kaviratna PD, Lan T, Pinnavaia TJ, (1994) Polym Prepr 35:788
83. Triantafillidis CS, LeBaron P, Pinnavaia TJ (2002) Chem Mater 14:4088
84. Triantafillidis CS, LeBaron P, Pinnavaia TJ (2002) J Sol State Chem 167:354
85. Winter H (1987) Polym Eng Sci 27:1698
86. Halley PJ, Mackay ME (1996) Polym Eng Sci 36:593
87. Halley PJ, Mackey ME, George GA (1994) High Perform Polym 6:405
88. Gillham JK (1986) Polym Eng Sci 26:1429
89. Bajaj P, Jha NK, Ananda Kumar R (1990) J Appl Polym Sci 40:203
90. Butzloff P, D'nouza NA, Golden TD, Garrett D (2001) Polym Eng Sci 41:1794
91. Xu W, Bao S, Shen S, Wang W, Hang G, He P (2003) J Polym Sci: Pt B: Polym Phys 41:378
92. Wang MS, Pinnavaia TJ (1994) Chem Mater 6:468
93. Becker O, Simon GP, Varley R, Halley P (2003) Polym Eng Sci 43:850
94. Gillham JK (1979) Polym Eng Sci 19:676
95. Aronhime MT, Gilham JK (1986) Adv Polym Sci 78:83
96. Ooi S, Cook WD, Simon GP, Such CH, (2000) Eur Polym J 41:3639
97. Xu WB, Bao SP, Shen SJ, Hang GP, He PS, (2003) J Appl Polym Sci 88:2932
98. Chen DZ, He PS, Pan LJ (2003) Polymer Testing 22:689
99. Weibing X, Pingsheng H, Dazhu C (2003) Eur Polym J 39:617
100. Dazhu C, Pingsheng H (2003) J Composite Mater 37:1275
101. Torre L, Frulloni E, Kenny JM, Manferti C, Camino G (2003) J Appl Polym Sci 90:2532
102. Lee A, Lichtenhan JD (1999) J Appl Polym Sci 73:1993
103. Kelly P, Akelah A, Qutubuddin S, Moet A (1994) J Mat Sci 29:2274
104. Chen KH, Yang SM (2002) J Appl Polym Sci 86:414
105. Shah AP, Gupta RK, Gangarao HVS, Powell CE (2002) Polym Eng Sci 42:1852
106. Ganguli S, Dean D, Jordan K, Price G, Vaia R (2003) Polymer 44: 6901
107. Lu H, Nutt S (2003) Macromolecules 36:4010
108. Zax DB, Yang DK, Santos RA, Hegemann H, Giannellis EP, Manias E (2000) J Chem Phys 112:2945
109. Kanapitsas A, Pissis P, Kotsilkova R (2002) J Non-Cryst Sol 305:204
110. Becker O, Varley RJ, Simon GP (2002) Opportunities for High Performance Epoxy Layered Silicate Nanocomposites. in Nanocomposites San Diego, California: ECM
111. Zerda AS, Lesser AJ (2001) J Polym Sci Part B: Polym Phys 39:1137
112. Zerda AS, Lesser AJ (2001) Intercalated clay nanocomposites: Morphology, mechanics and fracture behavior. Mat Res Soc Symp Vol 661, Nakatani AI, Hjelm RP, Gerspacher M, Krishnamoorti R, Warrendale MRS (eds), PA, USA
113. Kornmann X, Berglund LA, Sterte J (1998) Polym Eng Sci 38:1351
114. Becker O (2003) High Performance Epoxy-Layered Silicate Nanocomposites, PhD in School of Physics & Materials Engineering. Monash University: Melbourne
115. Basra C, Yilmazer U, Bayram G, (2003) ANTEC 2003 Proceedings, Society of Plastic Engineers p 3707

116. Isik I, Yilmazer U, Bayram G (2003) Polymer 44:6371
117. Gensler R, Gröppel P, Muhrer V, Müller N (2002) Part Syst Charact 19:293
118. Becker O, Varley RJ, Simon GP (2004) Eur Polym J 40: 187
119. Porter D, Metcalfe E, Thomas MJK (2000) Fire Mater 24:45
120. Gilman JW, Kashiwagi T, Lichtenhan JD (1997) SAMPE Journal 33:40
121. Gilman J, Kashiwagi T, Brown J, Lomakin S, Gianelis E (1998) Flammability Studies of Polymer Layered Silicate Nanocomposites. In 43rd International SAMPE Symposium
122. Lee DC, Jang LW (1997) J Appl Polym Sci 68:1997
123. Gu A, Liang G (2003) Polym Degrad Stab 80:383
124. Xie W, Gao Z, Pan WP, Hunter D, Singh A, Vaia R (2001) Chem Mater 13:2979
125. Hussain M, Varley RJ, Mathys Z, Cheng Y-B, Simon GP (2003) J Appl Polym Sci, 90: 3696
126. Kotsilkova R, Petkova V, Pelovski Y (2001) J Therm Anal Calorim 64:591
127. Park SJ, Seo DI, Lee JR (2002) J Colloid Interface Sci 251:160
128. Burnside SD, Giannelis EP (1995) Chem Mater 7:1597
129. Gilman JW, Kashiwagi T (2000) Polymer-Layered Silicate Nanocomposites with Conventional Flame Retardants. Polymer-Clay Nanocomposites, eds. TJ Pinnavaia, GW Beall, Wiley, Chichester
130. Zhu J, Uhl FM, Morgan AB, Wilkie C (2001) Chem Mater 13:4649
131. Gilman JW (1999) Appl Clay Sci 15:31
132. Gilman JW, Harris R, Hunter D (1999) Cyanate Ester Clay Nanocomposites: Synthesis and Flammability Studies. in Evolving and Revolutionary Technologies for the new Millenium – International SAMPE Symposium. Long Beach, CA
133. Gilman JW, Kashiwagi T, Nyden M, Brown JET, Jackson CL, Lomakin S, Giannelis EP, Manias E (1999) Flammability Studies of Polymer Layered Silicate Nanocomposites: Polyolefin, Epoxy, and Vinyl Ester Resins, in Chemistry and Technology of Polymer Additives, Ak-Malaika S, Golovoy A, Wilkie CA (eds), Blackwell Science Inc
134. Becker O, Varley RJ, Simon GP (2002) High Performance Epoxy Layered Silicate Nanocomposites: An Overview. In Materials Week. Munich, Germany
135. Becker O, Varley RJ, Simon GP (2003) J Mat Sci Lett 22:1411
136. Timmermann JF, Hayes BS, Seferis JC (2002) Composite Sci Technol 62:1249
137. Rice BP, Chen C, Cloos L, Curliss D (2001) SAMPE 37:7
138. Lelarge L, Becker O, Simon GP, Rieckmann T (2003) J Polym Sci Pt B, Polym Phys, unpublished data
139. Ratna D, Becker O, Krishnamurthy R, Simon GP, Varley R (2003) Polymer 44:7449
140. Frölich J, Thomann R, Mülhaupt (2003) Macromolecules 36:7205
141. Venderbosch RW, Meijer HEH, Lemstra PJ (1994) Polymer 35:4349
142. Venderbosch RW, Meijer HEH, Lemstra PJ (1995) Polymer 36:2903
143. Meijer HEH, Venderbosch RW, Goosens JPG, Lemstra PJ (1996) High Perf Polymer 8:133
144. Venderbosch RW, Meijer HEH, Lemstra PJ (1995) Polymer 36:1167
145. Frisch HL, Du Y, Schulz M, (1998) Interpenetrating Polymer Network (IPN) Materials in Polymer Networks – Principles of their formation structure and properties. Blackie Academic, Glasgow
146. Karger-Kocsis J, Gryschuk O, Fröhlich J, Müllhaupt R (2003) Composites Sci Technol 63:2045
147. Karger-Koscis J, Gryschchuk O, Jost N (2003) J. App Polym Sci 88:2124
148. Xie W, Xie RC, Pan WP, Hunter D, Koene B, Tan LS, Vaia R (2002) Chem Mater 14:4837

149. Vaia RA, Liu WD (2002) J Polym Sci Pt B Polym Phys 40:1590
150. Fornes TD, Paul DR (2003) Polymer 44:4993
151. Zilg C, Reichert P, Dietsche F, Engelhardt T, Mülhaupt R (1998) Kunststoffe 88:1812

Editor: Karel Dusek

Adv Polym Sci (2005) 179: 83–134
DOI 10.1007/b104480
© Springer-Verlag Berlin Heidelberg 2005
Published online: 6 June 2005

Proton-Exchanging Electrolyte Membranes Based on Aromatic Condensation Polymers

Alexandre L. Rusanov[1] (✉) · Dmitri Likhatchev[2] · Petr V. Kostoglodov[3] · Klaus Müllen[4] · Markus Klapper[4]

[1] A.N. Nesmeyanov Institute of Organoelement Compounds,
Russian Academy of Science, 28 Vavilova str., 119334 Moscow, Russia
alrus@ineos.ac.ru

[2] Materials Research Institute, UNAM, Cirquito Exterioir s/n, CU,
Apdo Postal 70-360 Coyoacan, 04510 Mexico City, Mexico
likhach@servidor.unam.mx

[3] YUKOS Research & Development Centre, 55/1 b.2, Leninski pr., 119333 Moscow, Russia
KostoglodovPV@yukos-rd.ru

[4] Max-Planck-Institut fur Polymerforschung, Ackermannweg 10, 55128 Mainz, Germany
meulen@mpip-mainz.mpg.de, klapper@mpip-mainz.mpg.de

1	Introduction	85
2	**Sulfonated Aromatic Condensation Polymers**	88
2.1	Sulfonation of High Molecular Mass Aromatic Condensation Polymers	88
2.2	Synthesis of Aromatic Condensation Polymers Based On Sulfonated Monomers	94
2.3	Properties of Sulfonated Aromatic Condensation Polymers	102
3	**Alkylsulfonated Aromatic Condensation Polymers and Proton-Conducting Electrolyte Membranes on their Basis**	109
4	**Proton-Exchanging Electrolyte Membranes Based On Polymer Complexes**	120
5	**Fuel Cell Applications of Proton-Exchanging Membranes Based On Aromatic Condensation Polymers**	127
6	**Conclusions**	128
	References	128

Abstract The results of the research and development of novel proton-exchanging membranes based on aromatic condensation polymers have been analysed and summarized with respect to their application in fuel cells. Primary attention has been paid to the basic properties of the starting polymers, such as thermal stability, water uptake and proton conductivity. General approaches to the preparation of aromatic condensation polymers with high proton conductivity have been considered, including direct suffocation, synthesis from monomers containing sulfonic acid groups, incorporation of alkylsulfonated substituents and formation of acid-basic polymer complexes. The bibliography includes 200 references.

Keywords Proton exchanging membranes · Fuel cells · Condensation polymers · Polyelectrolytes

Abbreviations

^{13}C-NMR	Carbon-13 nuclear magnetic resonance
^{1}H-NMR	Proton nuclear magnetic resonance
ACPs	Aromatic condensation polymers
DMAA	N,N-Dimethylacetamide
DMF	Dimethylformamide
DMSO	Dimethylsulfoxide
DNTA	Naphthalene-1,4,5,8-tetracarboxylic acid dianhydride
DSC	Differential scanning calorimetry
FT-IR	Fourier transform infrared spectroscopy
IR	Infrared spectroscopy
m-disulfo PBT	Poly[(benzo[1,2-*d*:4,5-*d'*]bisthiazole-2,6-diyl)-4,6-disulfo-1,3-phenylene]
m-sulfo PBT	Poly[(benzo[1,2-*d*:4,5-*d'*]bisthiazole-2,6-diyl)-5-sulfo-1,3-phenylene]
N-MP	N-Methyl-2-pyrrolidone
NMR	Nuclear magnetic resonance
PBI-MPS	(Methyl)propylsulfonated poly(benzimidazole)
PBI-PS	Propylsulfonated poly(benzimidazole)
PBP	Poly(4-benzoyl)-1,4-phenylene
PBTs	Poly(benzobisthiazoles)
PEEK	Poly(ether ether ketones)
PEMFS	Proton-exchanging membrane fuel cells
PES	Poly(ether sulfone)
PMFC	Polymer membrane fuel cell
PPA	Polyphosphoric acid
PPBP	Poly(4-phenoxybenzoyl-1,4-phenylene)
PPTA	Poly(*p*-phenylene terephthalamide)
PSPPI	Phenoxy substituted polyperyleneimide
PSSA	Poly(styrenesulfonic acid)
p-sulfo PBT	Poly[(benzo[1,2-*d*:4,5-*d'*]bisthiazole-2,6-diyl)-2-sulfo-1,4-phenylene]
S-PBI	Arylsulfonated poly(benzimidazole)
S-PEEK	Sulfonated poly(ether ether ketone)
S-PEES	Sulfonated poly(ether ether sulfone)
SPEFC	Solid polymer electrolytes fuel cells
S-PPBP	Sulfonated poly(4-phenoxybenzoyl-1,4-phenylene)
S-PPO	Sulfonated poly(phenylene oxide)
S-PPQ	Sulfonated polyphenylquinoxaline
S-PPS	Sulfonated poly(phenylene sulfide)
S-PPX	Sulfonated poly(p-xylylene)
TGA	Thermogravimetric analysis

1
Introduction

Proton-exchanging membrane fuel cells (PEMFC) are considered to be one of the most promising types of electrochemical device for power generation [1–10]. Low operation temperatures and the wide range of power make them attractive for portable, automotive, and stationary applications. However, advances made in these markets require further cost reduction and improved reliability. These can be achieved through development and implementation of novel proton-exchange membranes with higher performance and lower cost as compared to the state of the art polymeric electrolytes.

The basic design of a mono PEMFC cell is shown schematically in Fig. 1. The polyelectrolyte membrane is sandwiched between two noncorrosive porous electrodes. The electrochemical reactions occurring at the electrodes are the following:

on the anode: $\quad H_2 \rightarrow 2H^+ + 2e^-$;

on the cathode: $\quad 0.5O_2 + 2H^+ + 2e^- \rightarrow H_2O$;

net reaction: $\quad H_2 + 0.5O_2 \rightarrow H_2O + Q_1 + Q_2$,

where Q_1 is the electrical energy and Q_2 is the heat energy. Individual membrane electrode assemblies can be arranged into stacks to give the power range desired.

The proton-exchanging membrane is the most important component of the PEMFC. It must possess some specific properties [9], such as:

- a high ion-exchange capacity sufficient to provide a conductivity of the magnitude of $0.1\,S\,cm^{-1}$ at operational temperatures;

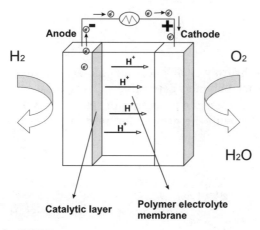

Fig. 1 A scheme of a PEMFC

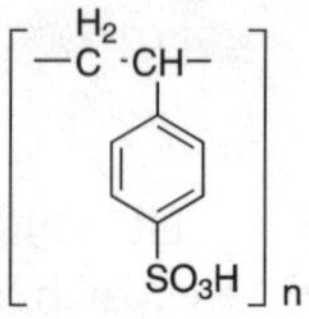

$$\left[\begin{array}{c} \underset{H_2}{-C}-CH- \\ \\ \\ SO_3H \end{array}\right]_n$$

Fig. 2

- low permeability to the fuel (hydrogen or methanol) and oxidant (oxygen) to prevent crossover of the membrane;
- substantial water uptake and good swelling to provide efficient dissociation of acids and to form a hydrated ionic phase within the entire volume of the membrane;
- sufficient chemical and mechanical stability for long-term operation under severe conditions (over 5000 h for electric vehicle applications).

The polymer membrane made of poly(styrenesulfonic acid) (PSSA) (Fig. 2) was used in the first PEMFC power plant built by General Electric in the mid-sixties for the Gemini space mission. The lifetime of these PMFCs was limited due to the degradation of the PSSA membrane under the impact of hydrogen peroxide radicals.

Further development and implementation of perfluorinated polymers [11–16] led to considerable advances in polymer electrolytes. The most widely-used fluorinated polymers are prepared by copolymerisation of tetrafluoroethylene with perfluorinated vinyl ethers of the following type

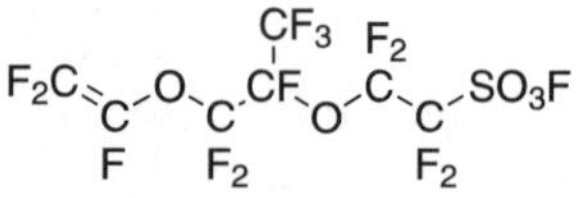

$$F_2C\underset{F}{\overset{}{\underset{\displaystyle C}{\diagdown}}}O\underset{F_2}{\overset{}{\underset{\displaystyle C}{\diagdown}}}\overset{CF_3}{\overset{|}{\underset{\displaystyle CF}{}}}O\underset{F_2}{\overset{F_2}{\underset{\displaystyle C}{\diagdown}}}\underset{F_2}{\overset{}{\underset{\displaystyle C}{\diagdown}}}SO_3F$$

Fig. 3

accompanied by hydrolysis of fluorosulfonic acid groups. Basic perfluorinated chains of such polymers determine high chemical and thermal stability, while side chains possess the properties of strong acids. Perfluorinated electrolyte membranes with the general formula shown below are also widely used.

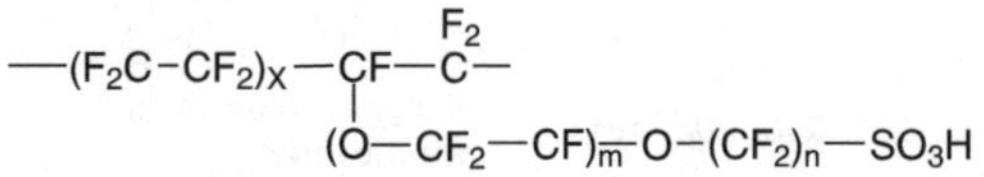

$$-(F_2C-CF_2)_x-\underset{\displaystyle (O-CF_2-CF)_{\overline{m}}O-(CF_2)_n-SO_3H}{\overset{F_2}{\underset{|}{CF-C}}}-$$

Fig. 4

Membrane	x	m	N
Nafion	6–10	1	2
Flemion	3–10	0.1	1–5
Aciplex-S	1.5–14	0.3	2–5
Dow membrane	3–10	0	2

The improved PMFC for the Gemini spacecraft was based on a perfluori-nated Nafion membrane. This membrane possesses substantially improved characteristics compared to the PSSA membranes; particular types of Nafion membranes are characterised by a lifetime of 50 000 h. Different types of Nafion membranes have different equivalent masses (grammes of polymer per mole H^+), namely, 1200 (Nafion 120), 1100 (Nafion 117 and Nafion 115) and 1000 (Nafion 105).

Perfluorinated membranes (Dow membrane) were developed by Dow Chemical Co. (USA). Their equivalent masses are equal to 800–850 g, while dry state thickness is of $\sim 5\,\mu m$. Flemion membranes with equivalent masses of ~ 1000 were developed by Asahi Glass Co. (Japan) [5]. Aciplex-S mem-branes were developed by Asahi Chemical Industry (Japan) and possess equivalent masses of 1000–1200 g.

All the membranes mentioned above, as well as Neosepta-F (Tokuyama, Japan) and Gore-Select (W L Gore and Associates Inc., USA) membranes possess a high proton conductivity (10^{-2}–10^{-1} S cm^{-1}) at water uptake up to 15 H_2O molecules per – SO_3H group and are characterised by good thermal, chemical and mechanical properties. On the other hand, these membranes are poor ionic conductors at reduced humidity and/or elevated temperatures. For instance, the conductivity of fully-hydrated Nafion membranes at room temperature reaches 10^{-2} S cm^{-1}. However, it dramatically decreases at 100 °C because of the loss of the absorbed water in the membranes. In addition, such membranes tend to undergo chemical degradation at elevated temperatures. Finally, their fabrication is rather expensive.

Therefore, the development of new solid polymer electrolytes, which com-bine sufficient electrochemical characteristics and low cost, is of current interest. A promising way of solving this problem involves preparation of membranes based on aromatic condensation polymers (ACPs). The chemistry of ACPs was characterised by considerable progress in the 1960–1990s [17–30].

ACPs have some advantages that make them particularly attractive:

- ACPs are cheaper than perfluorinated polymers and some of them are commercially available;
- ACPs containing polar groups have high water uptake over a wide tem-perature range;
- decomposition of ACPs can be to a great extent suppressed by proper molecular design;
- ACPs are easily recycled by conventional methods.

A number of reviews concerning the development of proton-conducting membranes based on polymer electrolytes are available [1, 7, 8, 31–33]. They contain information on the advanced materials, their electrochemical properties, water uptake and thermal stabilities. However, rapid accumulation of newly-obtained results gives an impetus to further generalisation of information in this field.

During the last decade, research on PEMFCs has been most intensively carried out in the following directions:

- development of sulfonated aromatic condensation polymers (ACPs) and membranes on their basis;
- development of alkylsulfonated ACPs and membranes on their basis;
- development of acid-basic polymer complexes and membranes on their basis.

2
Sulfonated Aromatic Condensation Polymers and Membranes On Their Basis

Aromatic polymers containing sulfonic acid groups can be prepared by sulfonation of high molecular mass ACPs or by condensation of monomers containing sulfonic acid groups.

2.1
Sulfonation of High Molecular Mass Aromatic Condensation Polymers

The simplest and the most widely-used method for the synthesis of sulfonated ACPs involves sulfonation of different classes of polymers, such as substituted poly-(1,4-phenylenes) [34, 35], poly-(p-xylylene [36, 37]), poly-(1,4-oxyphenylenes) [38–44], poly(ether ether ketones) (PEEK) [46–59], polyarylene(ether sulfones) [3, 60–74], poly(phenylene sulfides) [75], polyphenylquinoxalines [76–79], polybenzimidazoles [80], polyperyleneimides [81] and some other ACPs.

The chemical structures of sulfonated poly(4-phenoxybenzoyl-1,4-phenylene) (S-PPBP) (1), poly(p-xylylene) (S-PPX) (2), poly(phenylene sulfide) (S-PPS) (3), poly(phenylene oxide) (S-PPO) (4), poly(ether ether ketone) (S-PEEK) (5), poly(ether ether sulfone) (S-PEES) (6), arylsulfonated poly(benzimidazole) (S-PBI) (7) sulfonated polyphenylquinoxaline (S-PPQ) (8) and sulfonated phenoxy polyperyleneimide (PSPPI) (9) are shown below. ACPs are sulfonated using common sulfonating agents [82–85]. In particular, PEEK can be sulfonated in concentrated sulfuric acid [50], chlorosulfonic acid [86], SO$_3$ (either pure or as a mixture) [53, 65, 86, 87], a mixture of methanesulfonic acid with concentrated sulfuric acid [88] and acetyl sulfate [89, 90].

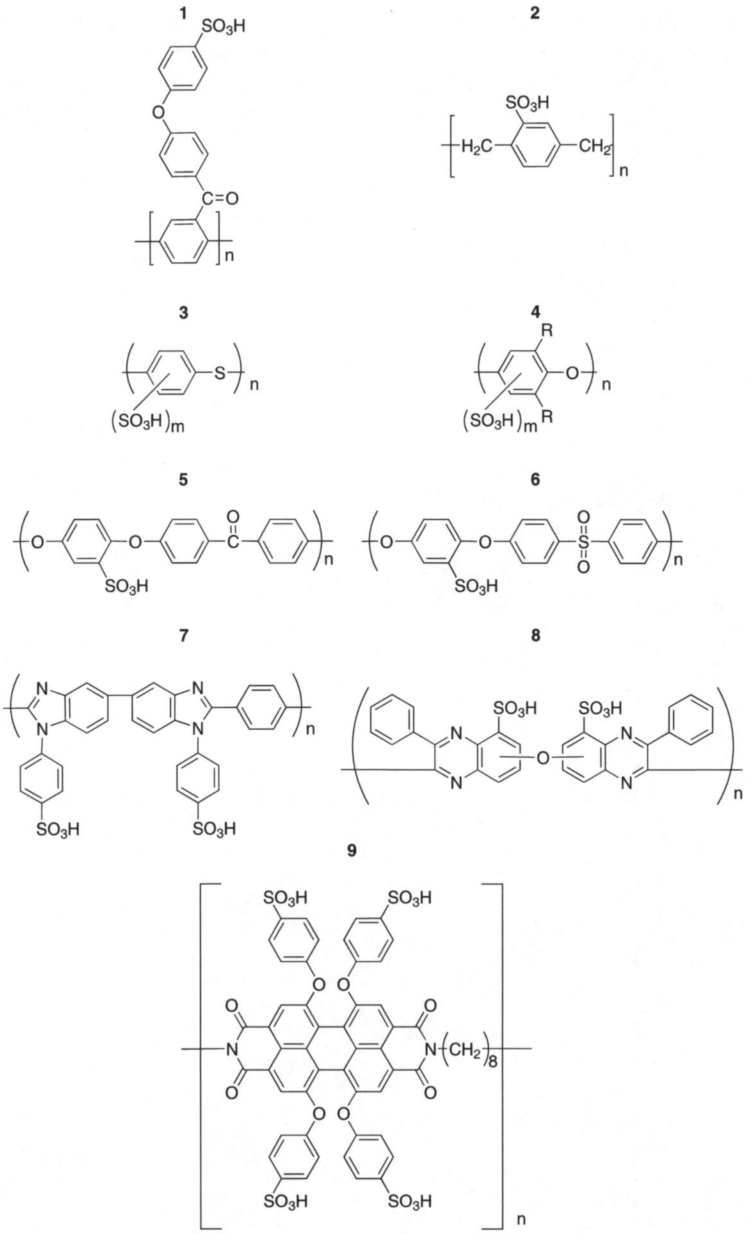

Fig. 5

Sulfonation of ACPs was systematically studied taking a number of polymers (first of all, PEEK and PPBP) as examples [7]. It was shown that sulfonation with chlorosulfonic or fuming sulfuric acid is sometimes accompanied by degradation of these polymers. The sulfonation rate of ACPs in sulfuric acid can be controlled by varying the reaction time and the acid concentration [91]. This technique allows preparation of target ACPs with sulfonation degrees ranging from 30% to 100% without chemical degradation or crosslinking of the polymer [92]. However, it should be noted that a direct sulfonation reaction cannot be used for preparation of random sulfonated copolymers and a sulfonation level of less than 30%, since sulfonation in sulfuric acid occurs under heterogeneous conditions due to high viscosity of the reaction solutions [49, 50]. For this reason, preparation of random copolymers requires the duration of the dissolution process to

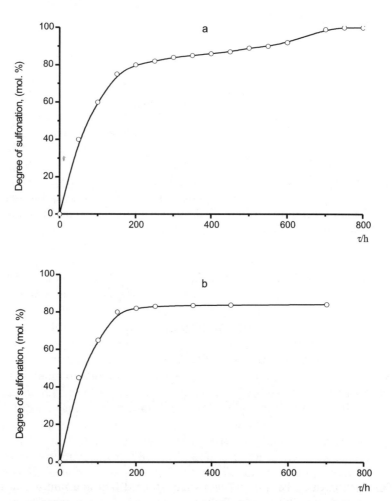

Fig. 6a,b Degree of sulfonation of PEEK (**a**) and PPBP (**b**) as a function of reaction time at room temperature [7, 35]

be shortened to 1 h. The dependences of the degree of sulfonation of PEEK and PPBP [35] on the reaction time at room temperature are shown in Fig. 6.

Sulfonation belongs to electrophilic substitution reactions, therefore, it strongly depends on the nature of substituents in the aromatic ring. Namely, electron-donating substituents favour the reaction whereas electron-withdrawing substituents slow it down significantly. For instance, in the case of PPBP, terminal phenyl rings in the side chain can be sulfonated under mild conditions similar to the sulfonation conditions for PEEK. In contrast to this, the phenyl ring substituent in poly(4-benzoyl)-1,4-phenylene (PBP), which contains an electron-withdrawing carbonyl group, cannot be sulfonated under these conditions [7]. The sulfonation level of PPBP and PEEK reaches nearly 80% within 100 h. The highest sulfonation degree for PPBP is 85%, whereas that of PEEK can be as high as 100%. This phenomenon can be attributed to steric hindrances to further sulfonation of PPBP in a viscous sulfuric acid solution.

The solubility of polymers changes while the degree of sulfonation increases. For instance, S-PEEK containing 30 mol% sulfonic acid groups dissolves in DMF, DMSO and N-methyl-2-pyrrolidone (N-MP); at 70% sulfonation the polymer is soluble in methanol, and at 100% – in water. Non-sulfonated PPBP is soluble in conventional chlorinated solvents (e.g., chloroform and dichloromethane), whereas S-PPBP with 30% sulfonation is insoluble in these solvents. However, the polymer can be dissolved in DMF, DMSO and N-MP. At the sulfonation levels above 65%, S-PPBP swells in methanol and water.

Sulfonation of PEEK in concentrated sulfuric acid at room temperature is accompanied by incorporation of not more than one sulfonic acid group into each repeating unit of the polymer [86, 90, 93, 94]. FT-IR spectroscopy studies show that PEEK is sulfonated at the phenylene ring between the ether groups.

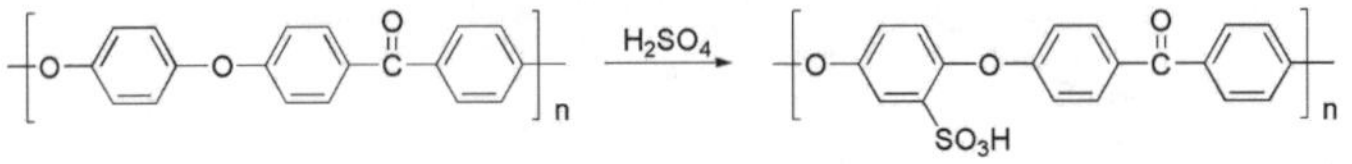

Scheme 1

Sulfonation of PPBP occurs at the p-position of the terminal phenoxy group.

Tsuchida et al. [95, 96] reported the synthesis of poly(thiophenylene sulfonic acid) containing up to two sulfonic acid groups per repeating unit. Polymerisation of 4-(methylsulfinyl)diphenyl sulfide in sulfuric acid upon heating or in the presence of SO_3 resulted in sulfonated poly(sulfonium cation), which was then converted into the corresponding sulfonated poly(phenylene sulfide).

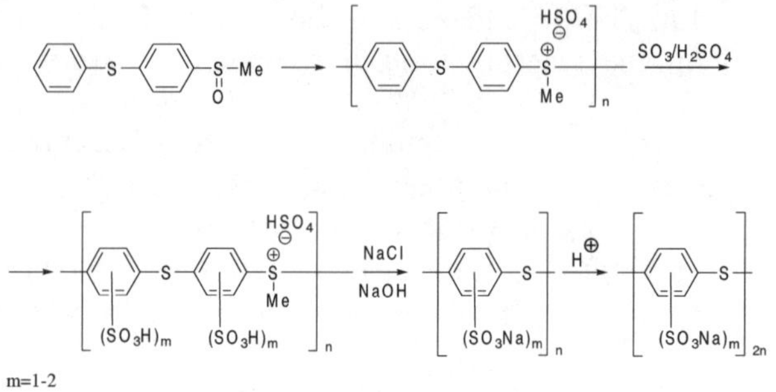

Scheme 2

The course of sulfonation was controlled by varying the reaction time, the temperature and/or by adding SO_3. Polymer electrolyte thus obtained is soluble in water and methanol and can form a transparent film.

Novel polymer electrolytes exhibiting high proton conductivity (higher than $10^{-2}\,S\,cm^{-1}$) were prepared by sulfonation of poly(ether sulfone) (PES) [97, 98]. In these polymers the protons of the sulfonic acid groups are partially replaced by metal ions (Mg, Ti, Al, Ln) which leads to extension of the durability of the electrolytes.

Sulfonated polyphenylquinoxalines were prepared using two approaches: sulfonation of polyphenylquinoxalines prepared by the conventional technique [76–79, 99], or synthesis of polyphenylquinoxalines directly in the sulfonating medium [76–79, 99]:

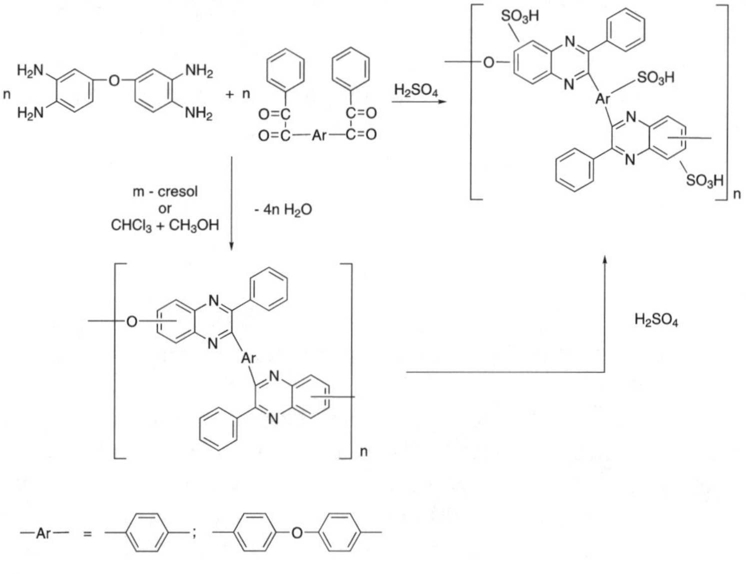

Scheme 3

In both cases, sulfonation was performed in sulfuric acid: oleum mixture (4 : 1) at 125 °C. High-strength thermally stable films showing high hydrolytic stability were cast from the solution of sulfonated polyphenylquinoxalines in DMF.

Another approach to the preparation of sulfonated polyphenylquinoxalines is based on the treatment of polyphenylquinoxalines containing activated fluorine atoms [100, 101] with hydroxyarylsulfonic acids [99, 102]:

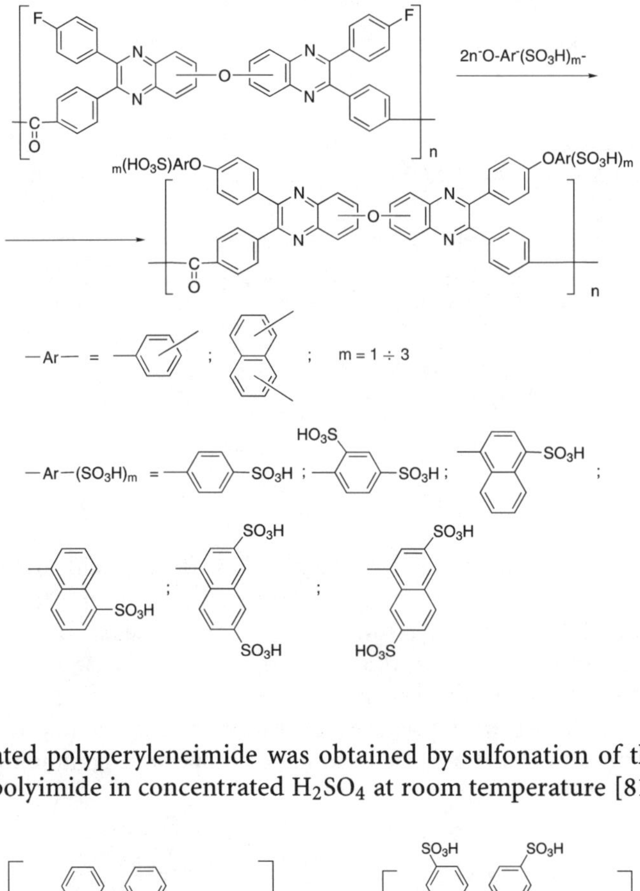

Scheme 4

Sulfonated polyperyleneimide was obtained by sulfonation of the corresponding polyimide in concentrated H_2SO_4 at room temperature [81].

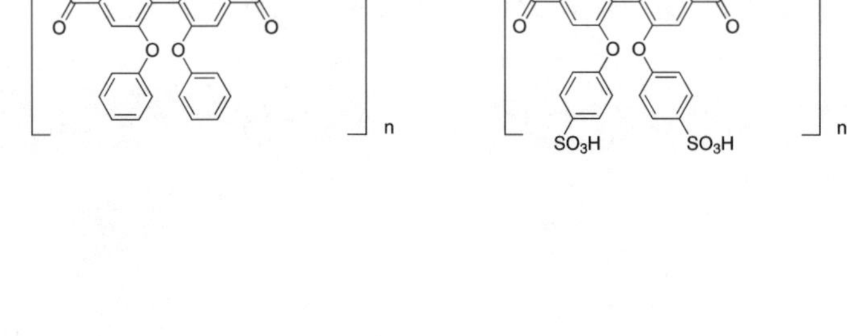

Scheme 5

No evidence for side reactions, e.g. three-fold sulfonation or cleavage of phenoxy substituents was detected by ^{1}H-NMR spectroscopy.

The organic solvent solubility of the starting polyimide can be modified by sulfonation to give water-soluble polyimide possessing film-forming properties.

2.2
Synthesis of Aromatic Condensation Polymers Based On Sulfonated Monomers

Sulfonated ACPs are prepared both by direct sulfonation and by polycondensation and polycyclocondensation of sulfonated compounds. Synthesis of the first sulfonated poly(p-phenylene) was reported by Wegner and co-workers [103]. The Suzuki coupling of diboronic ethers with dibromoaromatic monomers furnished poly(p-phenylene) with 95% yield.

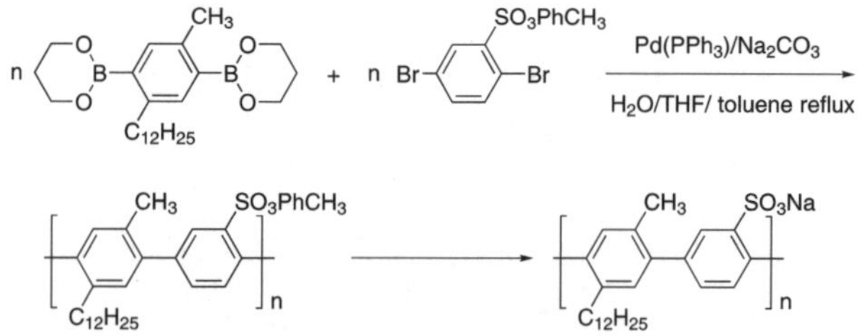

Scheme 6

Absolute molecular weight of 36 kD was determined by membrane osmometry in toluene. Surprisingly, the final polymer was not soluble in a basic aqueous solution, but it was soluble in DMSO. In a subsequent report from the same research group [104, 105] this synthetic approach was extended to produce other isomeric structures.

Sulfonated PEEKs were prepared by the reactions of sulfonated hydroquinone with difluoro-substituted aromatic compounds containing carbonyl groups [106, 107]:

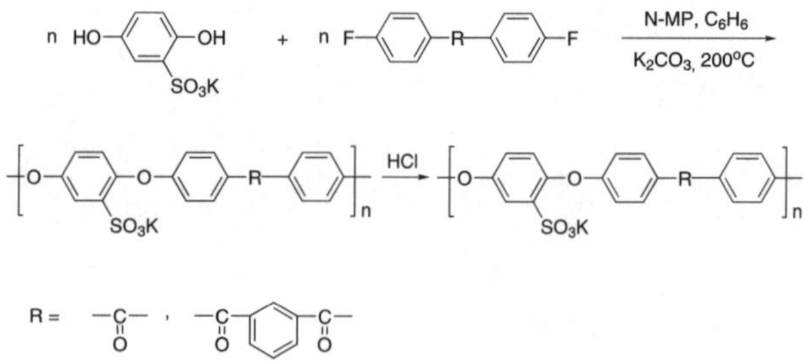

Scheme 7

Novel sulfonated PEKs were prepared directly by nucleophilic polycondensation of 4,4'-sulfonyldiphenol with various ratios of 4,4'-difluorobenzo-
phenone to 5,5'-carbonyl-bis-(2-fluorobenzenesulfonate) in DMSO [108].

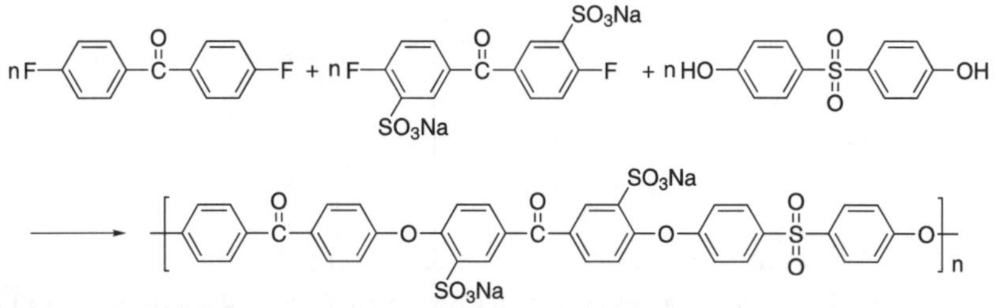

Scheme 8

The resulting polyelectrolytes have been characterised by IR, NMR, TGA
and DSC. The 10% weight loss temperature is higher than 510 °C, which
indicates that the introduction of 4,4'-sulfonyldiphenol with the powerful
electron-withdrawing group – SO_2 – into the main chain of sulfonated PEK
improves the thermal stability against desulfonation.

Sulfonated poly(phthalazinone ether ketones) were synthesized directly by
aromatic nucleophilic polycondensation of 4-(4-hydroxyphenyl)phthalazino-
ne with various ratios of 5,5'-carbonylbis-(2-fluorobenzenesulfonate) or 4,4'-
difluorobenzophenone [109].

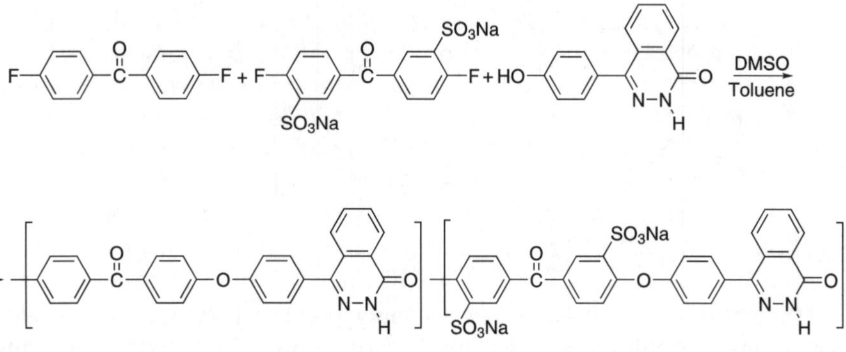

Scheme 9

The 10% weight loss temperature of the product is higher than 500 °C.

An analogous procedure was employed in recent studies [110–114] on
the synthesis of poly(arylene ether sulfones) using reactions of sulfonated
4,4-dichlorodiphenyl sulfone with various bisphenols.

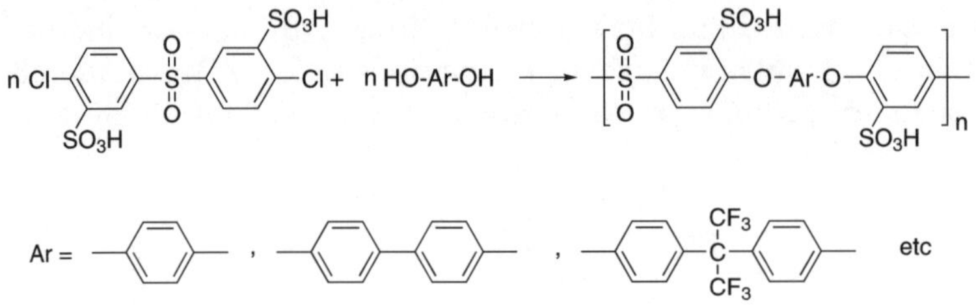

Scheme 10

The use of *m*-aminophenol as an additive along with bis-phenols allowed the preparation of poly(arylene ether sulfones) with terminal amino groups [115].

Sulfonated poly(thiophenylene sulfones) were prepared by the interaction of sulfonated 4,4′-difluorodiphenyl sulfone with 4,4′-dimercaptobenzophenone [116]. Not only homopolymers, but also copolymers were obtained. In the latter case, the fraction of sulfonated 4,4′-difluorodiphenyl sulfone was replaced with nonsulfonated monomers.

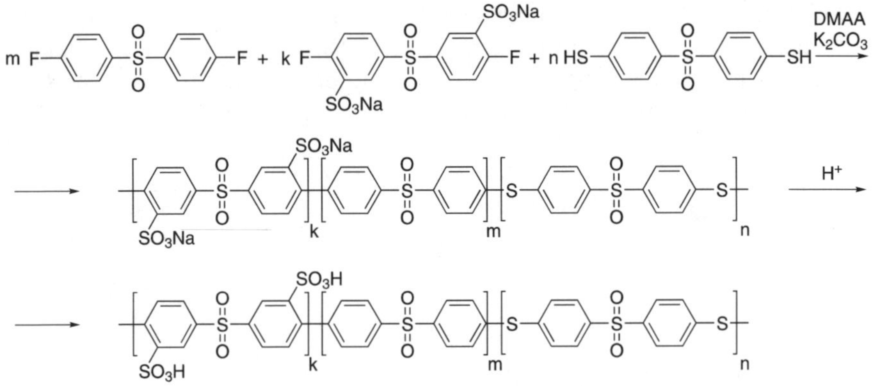

Scheme 11

Using this approach, one can not only prepare polymers with regular arrangement of sulfonic acid groups, but, sometimes, introduce a large number of sulfonic acid groups into the ACP macromolecules compared to the sulfonation of ACPs.

Sulfonated poly(phthalazinone ether sulfones) were directly prepared by polycondensation of 4-(4-hydroxyphenyl)phthalazinone with various rations of disodium salt of 5,5′-sulfonylbis-(2-fluorobenzenesulfonate) to 4-fluorophenylsulfone [117].

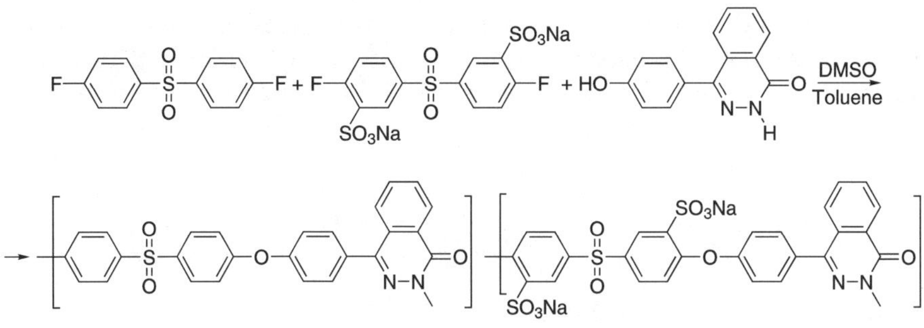

Scheme 12

The resulting ionomers demonstrated high molecular weight, high ion-exchange capacity and low swelling. Low swelling originates from the hydrogen bonding between hydrogen atoms of sulfonic acid and carbonyl groups, which has been validated by variable temperature IR spectra.

High molecular weight water soluble sulfonated polyamides were prepared by the interaction of sulfonated diamines with terephthalic and isophthalic acid dichlorides [118–122].

For this purpose the following diamines were used:

1. 4,4′-diaminobiphenyl-2,2′-disulfonic acid (10)
2. 4,4′-diaminostilbene-2,2′-disulfonic acid (11)
3. para or metadiaminobenzene sulfonic acid (12)
4. 2,5-diaminobenzene-1,4-disulfonic acid (13)

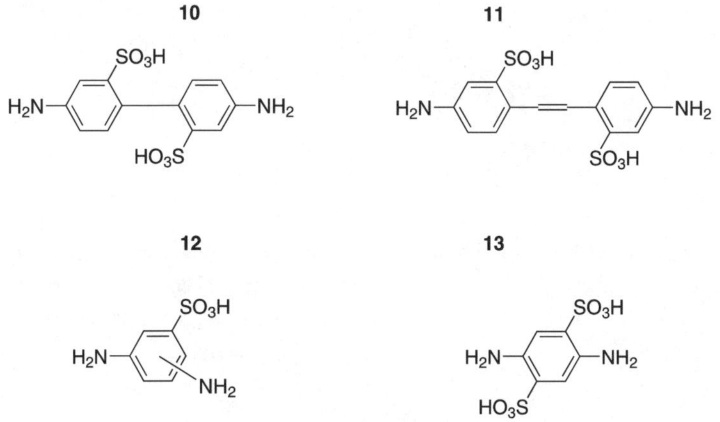

Fig. 7

Some polymers had a sufficiently high molecular weight (more than 200 000), extremely high intrinsic viscosity ($\sim$ 65 dl/g), and appeared to transform into a helical coil in saline solution.

Sulfonated monomers were also used for the synthesis of sulfonated poly-imides [123, 124]. In particular, sodium salt of the sulfonated bis-4-[(3-aminophenoxy)phenyl]phenylphosphine oxide was used for the preparation of sulfonated polyimides [123].

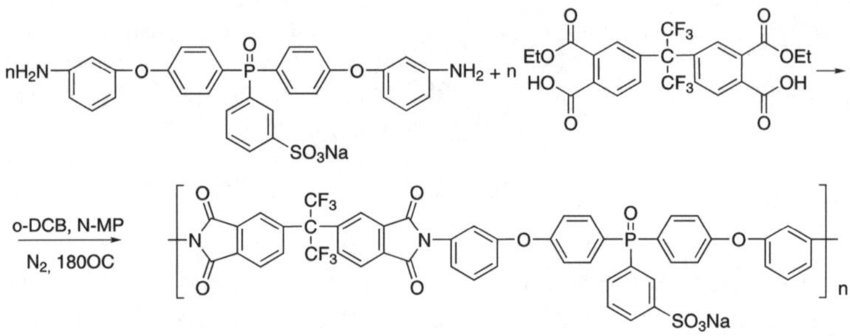

Scheme 13

Of particular interest is the use of 4,4'-diamino-2,2'-diphenylsulfonic acid [124–126] produced on a semi-industrial scale as a sulfonated monomer for preparation of polyimides. The reactions of a mixture of this monomer and 4,4'-diaminodiphenylmethane and 4,4'-diaminodiphenyloxide with di-phenyloxide-3,3',4,4'-tetracarboxylic acid dianhydride resulted in sulfonated polyimides [124] with the following structure:

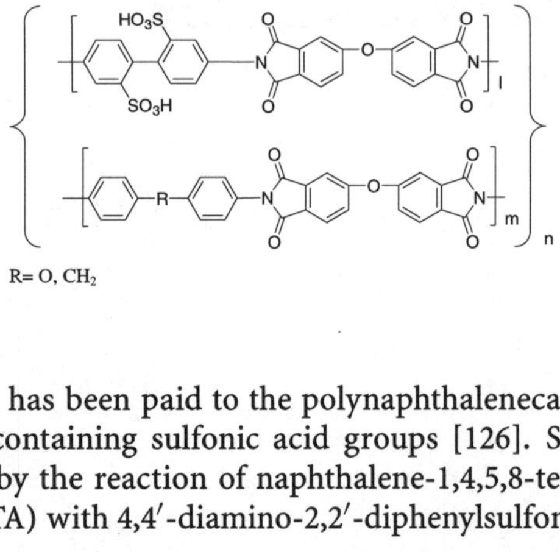

R= O, CH$_2$

Fig. 8

Great attention has been paid to the polynaphthalenecarboximides (poly-naphthylimides) containing sulfonic acid groups [126]. Such polymers are usually prepared by the reaction of naphthalene-1,4,5,8-tetracarboxylic acid dianhydride (DNTA) with 4,4'-diamino-2,2'-diphenylsulfonic acid

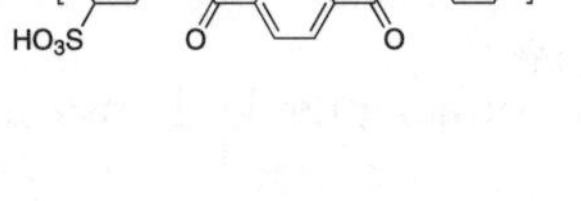

Fig. 9

or a mixture of this sulfonated monomer with other aromatic diamines.

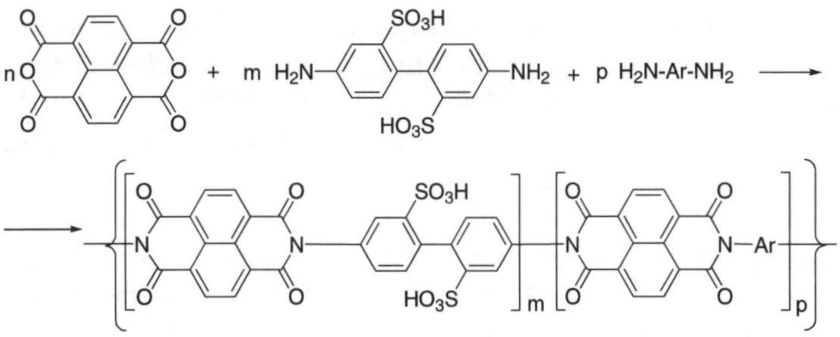

n=m+p

Scheme 14

Almost all studies on the synthesis of poly(naphthylimides) based on 4,4'-diamino-2,2'-diphenylsulfonic acid were aimed at preparing copolymers with controlled properties that could be varied over a wide range.

Other diamines were used for a similar purpose and these were the following:

1. 4,4'-diaminodiphenylamino-2-sulfonic acid (14) [138];
2. sulfonated bis-(3-aminophenyl)phenyl phosphine oxide (15) [136];
3. 3,3-disulfonate-bis[4-(3-aminophenoxy)phenyl]sulfone (16) [137];
4. 9,9-bis(4-aminophenyl)fluorene-2,7-disulfonic acid (17) [139–142];
5. 4,4'-diaminodiphenyl ether-2,2'-disulfonic acid (18) [142, 143];
6. 4,4'-bis-(4-aminophenoxy)biphenyl-3,3'-disulfonic acid (19) [142, 143].

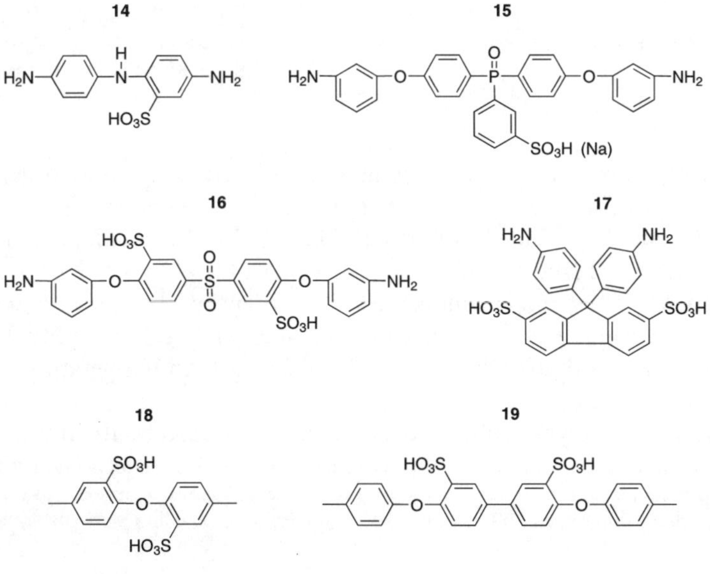

Fig. 10

In general, poly(naphthylimides) containing six-membered imide rings in backbones are characterised by significantly improved chemical resistance compared to analogous poly(phthalimides) [144–147]. A similar conclusion was made comparing the chemical resistance of sulfonated polyimides and poly(naphthylimides) [124, 125].

Several attempts have been made to develop sulfonated polyazoles [148, 149] and polybenzazoles [150–160]. Sulfonated poly-1,3,4-oxadiazoles have been prepared by the interaction of 5-sulfoisophthalic acid with hydrazine sulfate in polyphosphoric acid (PPA) [148, 149]

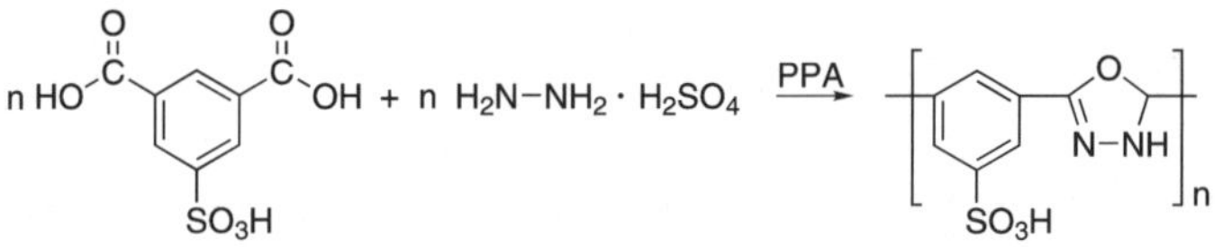

Scheme 15

Sulfonated polybenzimidazoles have been prepared by polycondensation of sulfoterephthalic acid and disulfoisophthalic acid with 3,3'-diaminobenzidine using high temperature solution polycondensation in PPA [150–153].

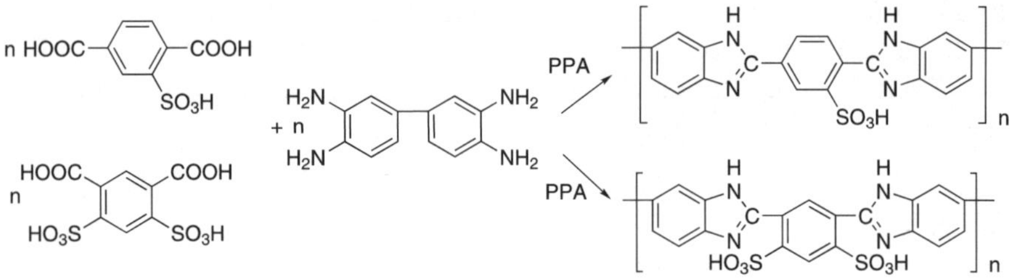

Scheme 16

The polymers obtained were soluble in sulfuric acid, some organic solvents, and aqueous strong alkaline solutions. The polymers were stable up to 400 °C, but they yielded polybenzimidazoles by eliminating sulfonic acid groups, instead of ring closure.

Sulfonated polybenzobisimidazoles were prepared by the interaction of 1,2,4,5-tetraaminobenzene tetrahydrochloride with sulfoterephthalic acid [154, 155] and 5-sulfoisophthalic acid [156] using high temperature solution polycondensation in PPA.

Alternatively polybenzobisimidazoles were prepared by the interaction of 1,2,4,5-tetraaminobenzene tetrahydrochloride with 4-carboxy-2-sulfobenzoic anhydride [157]:

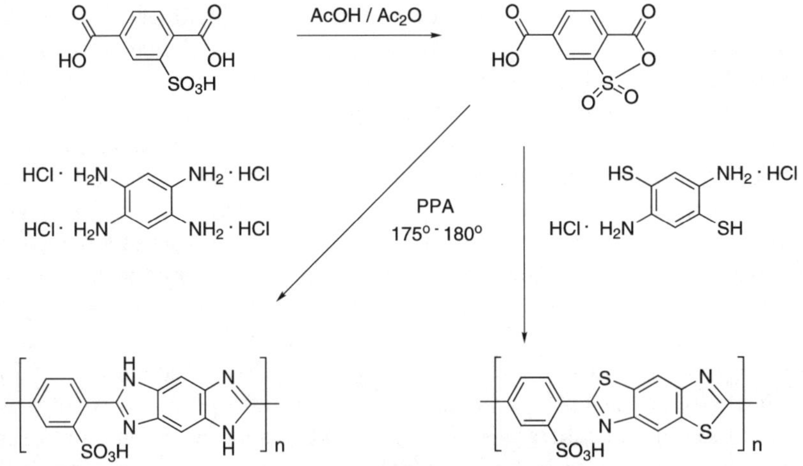

Scheme 17

The same synthetic approach was used for the preparation of sulfonated polybenzobisthiazoles [157] (Scheme 17).

Aromatic polyelectrolytes based on sulfonated poly(benzobisthiazoles) (PBTs) have been synthesized also by polycondensation of sulfo-containing aromatic dicarboxylic acids with 2,5-diamino-1,4-benzenedithiol dihydrochloride (DABDT) in freshly prepared PPA [158].

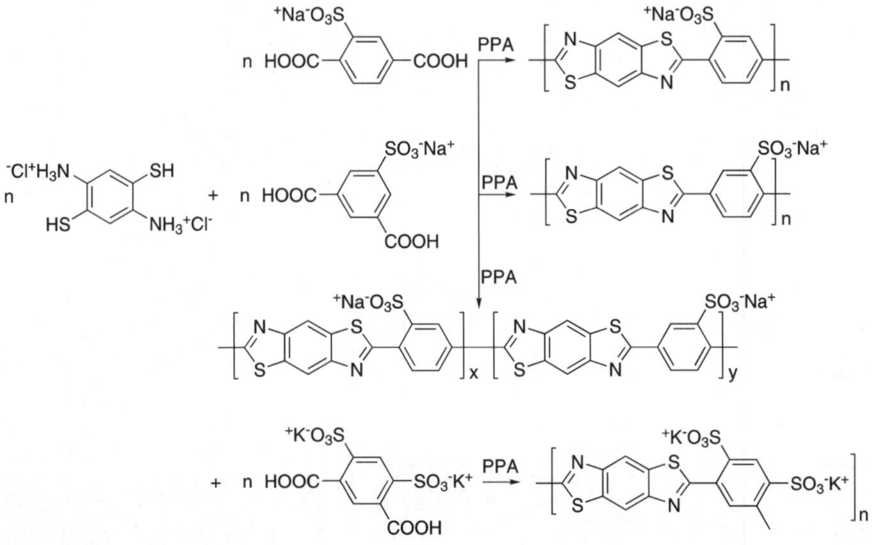

Scheme 18

Several sulfonated PBTs, poly[(benzo[1,2-*d*:4,5-*d'*]bisthiazole-2,6-diyl)-2-sulfo-1,4-phenylene] sodium salt (p-sulfo PBT), poly[(benzo[1,2-*d*:4,5-*d'*]-bisthiazole-2,6-diyl)-5-sulfo-1,3-phenylene] sodium salt (m-sulfo PBT), their copolymers, and poly[(benzo[1,2-*d*:4,5-*d'*]bisthiazole-2,6-diyl)-4,6-disulfo-1,3-phenylene] potassium salt (m-disulfo PBT), have been targeted and the polymers obtained characterised by ^{13}C-NMR, FT-IR, elemental analysis, thermal analysis and solution viscosity measurements. Structural analysis confirms the structure of p-sulfo PBT and m-disulfo PBT, but suggests that the sulfonate is cleaved from the chain during synthesis of m-disulfo PBT. The polymer m-disulfo PBT dissolves in water as well as strong acids, while p-sulfo PBT dissolves well in strong acids, certain solvent mixtures containing strong acids, and hot DMSO. TGA indicates that these sulfonated PBTs are thermally stable up to over 500 °C. Free-standing films of p-sulfo PBT, cast from dilute neutral DMSO solutions, are transparent, tough, and orange in colour. Films cast from basic DMSO are also free standing, while being opaque yellow-green.

Sulfonated polybenzoxazoles were prepared from 5-sulfoisophthalic or 2-sulfoterephthalic acids and different bis-(o-aminophenols) [159].

Disulfonated polybenzoxazoles were prepared by the interaction of 2,2'-bis-(3-amino-4-hydroxy-phenyl)hexafluoropropane with disodium-2,2'-disulfonate-4,4'-oxydibenzoic acid and 4,4'-oxydibenzoic acid using PPA as the polymerisation media [160].

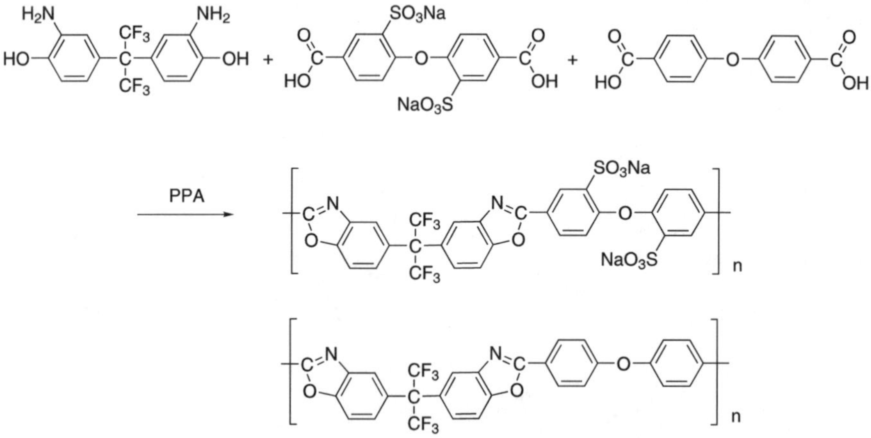

Scheme 19

2.3
Properties of Sulfonated Aromatic Condensation Polymers

The most important properties of sulfonated ACPs are their thermal stability, water uptake and proton conductivity. PEMFCs and electrochemical devices on their basis operating in a temperature range of 100–200 °C require polymer electrolyte membranes characterised by fast proton transfer. The operation of PEMFCs at elevated temperature has a number of advantages. It causes an increase in the rates of fuel cell reactions and reduces catalyst poisoning with absorbed carbon monoxide, thus reducing the demand for catalysts.

Thermal stability of polymer membranes based on S-PPBP has been studied [7, 35] by sample heating followed by elemental analysis (thermogravimetric analysis, or TGA, at a heating rate of 10 °C min^{-1} under nitrogen) (Fig. 11).

According to the results of TGA studies, S-PPBP showed a mass loss of nearly 20% in the temperature range between 250 and 400 °C, which corresponds to the decomposition of sulfonic acid groups.

The dependence of the degradation temperature, (T_d), of S-PPBP and S-PEEK on the degree of sulfonation is presented in Fig. 12.

Degradation of sulfonated polymers was observed between 250 °C and 350 °C, i.e., at temperatures that are much lower then those for non-sulfonated PPBP and PEEK.

As the degree of sulfonation increased, the degradation temperatures decreased from 500 down to 300 °C for S-PEEK, and from 500 down to 250 °C for S-PPBP. The results of elemental analysis of residues indicate a dramatic (nearly ten-fold) decrease in sulfur content of the polymers after heating at temperature above 400 °C. These data confirm that thermal stabilities of polymers are sufficient for fuel cell application even at high sulfonation levels [7, 35].

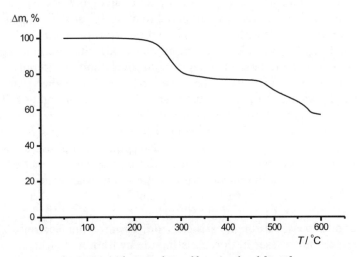

Fig. 11 TGA curve of S-PPBP with 80 mol % sulfonation level [7, 35]

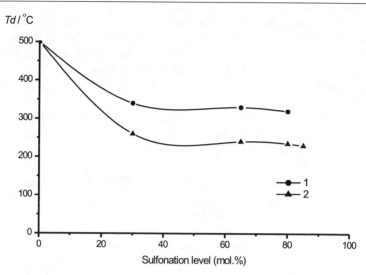

Fig. 12 Degradation temperature of S-PEEK (1) and S-PPBP(2) as a function of sulfonation level [7, 35]

Other proton-conducting polymer electrolytes based on sulfonated aromatic condensation polymers also show the onset of thermal degradation at temperatures between 200 and 400 °C. Desulfonation of arylsulfonic acids occurs readily upon heating their aqueous solution up to 100–175 °C. Therefore, desulfonation imposes limitations on the thermal stability of sulfonated aromatic condensation electrolytes. It should be mentioned that the presence of bulky substituents attached to the phenyl rings can, to some extent, favour an increase in the onset of thermal degradation temperature.

According to Tsuchida et al. [96], highly sulfonated poly(phenylene sulfide) exhibits higher thermal stability compared to other sulfonated aromatic polymer electrolytes. This conclusion was based on the results of a TGA study of thermal stability of poly(thiophenylenesulfonic acid) with different degrees of sulfonation. The degradation temperature of highly sulfonated polymer (degree of sulfonation $m = 2, 0$) is 265 °C, which is 125 °C higher than that of the low sulfonated polymer ($m = 0, 6$). The C – S bond in highly sulfonated polymer is stronger due to the presence of two electron-withdrawing sulfonic acid substituents in each benzene ring. The initial mass loss of this polymer at 265–380 °C is only 13%, which correspond to the loss of two H_2O molecules per repeating unit. Therefore, the desulfonation reaction in this polymer slows down upon introduction of the electron-acceptor.

Water is carried into the fuel cell with humidified gas (H_2, O_2) steams and enters electrodes as a result of gas diffusion. A mixture of liquid water and water vapours passes through each electrode towards the electrode/electrolyte interface and crosses it, thus assisting the hydration of electrolyte membranes. Oxygen reduction at the cathode provides an additional source of water.

Water transport through the membrane occurs due to electro-osmotic drag of water by proton transfer from anode to cathode and due to diffusion of water molecules across concentration gradients.

Optimum hydration level of electrolyte membranes is a key factor for normal fuel cell operation. If the electrolyte membrane is too dry its conductivity decreases, whereas an excess of water in the membrane can lead to cathode flooding. In both cases fuel cell performance drops.

Absorbtion of water vapour by polymer films prepared from S-PEEK and S-PPBP was studied by placing films into the atmosphere with different humidities and subsequent measuring of the equilibrium water content. The results obtained were found to be close to those reported in similar studies for Nafion membranes [10]. The dependence of water uptake for S-PEEK and S-PPBP films on relative humidity at room temperature is shown in Fig. 13.

Assuming the water activity and water content in the membrane obey Raoult's law, the activity coefficient of water in the polymer is larger than unity at relative humidities exceeding a particular value. The equilibrium content of water in S-PEEK and S-PPBP increases as the sulfonation level increases. At relative humidities in the range from 0% to 50% (first region) a relatively small increase in the water uptake is observed, whereas an increase in the relative humidity from 50% to 100% (second region) leads to a much greater increase in the water uptake. The first region corresponds to water uptake due to solvation of the proton and sulfonate ions. During

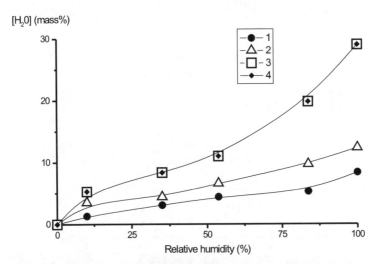

Fig. 13 Water uptake of S-PPBP (1–3) and S-PEEK (4) at room temperature as a function of relative humidity [7]. Concentration of SO_3H groups in the polymer (mol%): 30 (**1**), 65 (**2**), 80 (**3**) and 65 (**4**)

solvation, water is involved in the interaction with ionic components of the polymer. These interactions overcome the tendency of the polymer to exclude water due to its hydrophobic nature and resistance to swelling [7]. The second region corresponds to the uptake of water involved in polymer swelling.

The content of water in S-PPBP at 65 mol % sulfonation is higher than for S-PEEK with the same sulfonation level. At a relative humidity of 100% and room temperature, the content of water in S-PPBP and S-PEEK is 8.7 and 2.5 molecules per sulfonic acid group, respectively.

Picnometric measurements showed that the densities of the polymers with a sulfonation level of 65 mol % were 1.338 (S-PEEK) and 1.373 g cm^{-3} (S-PPBP). According to the results obtained by scanning electron microscopy, both polymers exhibited very close characteristics of their surface and fracture surface.

The difference in water uptake between S-PEEK and S-PPBP can be attributed to flexibility of the phenoxybenzoyl group in the side chain of S-PPBP, which favours water permeation into the polymer and water absorption by the terminal sulfonic acid group. Water uptake of S-PPBP is comparable to that of Nafion membranes.

DTA studies revealed a rather strong interaction between water molecules in sulfonated hydrocarbon polymers and their sulfonic acid groups, which leads to high proton conductivities at high temperature and low humidity.

Proton conductivity of sulfonated poly(phenylene sulfide) is 10^{-5} S cm^{-12} at room temperature and relative humidity of 30%. The conductivity exponentially grows with the increase in relative humidity and reaches a value of 2×10^{-2} cm^{-1} at 94% humidity (Fig. 14).

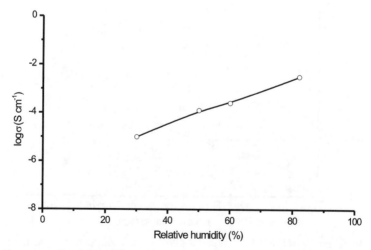

Fig. 14 Proton conductivity of the sulfonated polyphenylene sulfide (Scheme 2) ($m = 2$) at room temperature as a function of relative humidity [95]

In this case, the content of water in the polymer is 10.3 molecules per sulfonic acid group. The maximum conductivity of sulfonated poly(phenylene sulfide) ($m = 2.0$) at 80 °C was 4.5×10^{-2} S cm^{-1}.

Experiments [7] on water absorption by S-PEEK and S-PPBP films showed that proton conductivities of the films containing equilibrium amounts of absorbed water depend on the relative humidity. Fig. 15 represents the dependency of the proton conductivities of S-PEEK and S-PPBP with different sulfonation levels as a function of relative humidity.

It becomes clear that proton conductivities of the films increase with the relative humidity and water uptake and can become as high as 10^{-5} S cm^{-1} (for S-PEEK).

The proton conductivities for S-PEEK and S-PPBP with equal degrees of sulfonation (65 mol %) at a 100% relative humidity can be compared using the graphs shown in Fig. 16.

It is obvious that the proton conductivities and water uptake for S-PPBP are much higher than those for S-PEEK. Moreover, the proton conductivity for S-PEEK dramatically decreases at temperature above 100 °C, whereas that of S-PPBP appears to be much less temperature dependent.

Sulfonated poly(phenylene sulfide) and S-PPBP exhibit stable proton conductivities at elevated temperatures. For this reason, they are considered as prospective polymers for manufacture of proton-conducting electrolyte membranes operating at elevated temperatures and low humidity.

On the contrary, the conductivity of perfluorinated polymer electrolytes usually appreciably decreases with increasing temperature, that is, the conductivity of such electrolytes at 80 °C is by an order of magnitude lower than

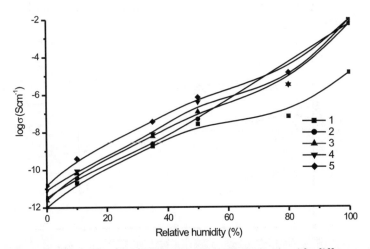

Fig. 15 Proton conductivity of S-PEEK (1) and S-PPBP (2–5) with different sulfonation levels as a function of relative humidity at room temperature [7]. Sulfonation level (mol %): 65 (1), 30 (2), 65 (3), 80 (4) and 85 (5)

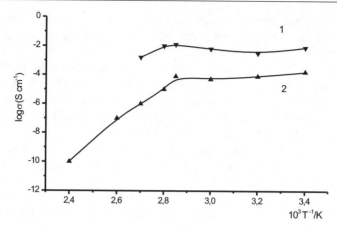

Fig. 16 Temperature dependences of proton conductivity as S-PPBP (1) and S-PEEK (2) with the same degrees of sulfonation (65 mol %) at a relative humidity of 120% [7]

at 60 °C. Perfluorinated polymer membranes become less conducting at high temperatures, since the loss of water causes the channels to collapse, thus making proton transport more difficult.

In particular, proton conductivity of Nafion membranes at temperatures above 100 °C dramatically decreases due to their dehydration.

Figure 17 represents temperature dependences of the proton conductivity of S-PEEK with a sulfonation degree of 85 mol % at different relative humidity values.

Similarly to Nafion, the proton conductivity of S-PEEK substantially drops as the humidity decreases [86]. The dependence of proton conductivity on

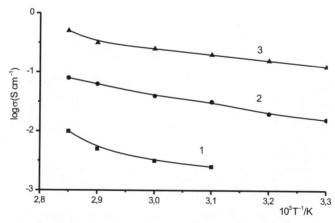

Fig. 17 Temperature dependence of proton conductivity of an S-PEEK membrane with sulfonation level of 85% at relative humidity of 50% (1), 70% (2) and 90% (3) [7]

humidity reflects a tendency of S-PEEK to absorb water vapours. This can be attributed to a "liquid" proton conductivity mechanism when protons are transported in the form of hydronium ions through water-filled pores of the membrane [32].

S-PEEK samples exhibit a slight increase in conductivity with temperature at all relative humidities (50%, 70% and 90%). This can be due to the strong interaction between the sulfonic acid groups and the absorbed water molecules.

Proton-conducting polymer electrolyte membranes based on ACPs such as S-PPBP and sulfonated poly(phenylene sulfide) contain rather large amounts of bound water. This seems to be the reason for such a salient feature of these membranes as an increased proton conductivity at high temperatures and/or low humidities. This conclusion was confirmed by the results of differential scanning calorimetry (DSC) studies of these systems [7].

3
Alkylsulfonated Aromatic Condensation Polymers and Proton-Conducting Electrolyte Membranes on their Basis

The major drawback of sulfonated proton-conducting polymer electrolytes is their degradation at 200–400 °C occurring due to desulfonation. By introducing alkylsulfonated substituents into the macromolecules of aromatic polymers one can prepare thermostable proton-conducting polymers. Their electrochemical properties can be controlled by varying the number of substituents and the length of alkyl chains. Water uptake and proton conductivity of alkylsulfonated polymers are close to those of sulfonated electrolytes that exhibit high thermal and chemical stability and mechanical strength.

Poly(p-phenyleneterephthalamido-N-propylsulfonate) and poly(p-phenyleneterephthalamido-N-benzylsulfonate) were synthesised using correspondent polyamides containing reactive NH groups [161]. The polyamides were modified by treatment with NaH in DMSO [161, 162], and the resulting polyanion obtained was introduced into the reaction with 1,3-propane sultone (Scheme 20).

A similar approach was employed for the modification of poly(benzimidazoles) (PBI) [163–171].

Yet another synthetic route to obtaining sulfonated PBI involves treatment of the above mentioned polyanion with 4-bromobenzyl sulfonate resulting in poly[2,2'-m-phenylene-bi(N-benzylsulfonate)benzimidazolo-5,5'-diyl] (Scheme 21).

Compared to starting polymers, alkylsulfonated PBI is more soluble in polar organic solvents (DMAA or DMSO). The solubility depends on the degree of alkylsulfonation.

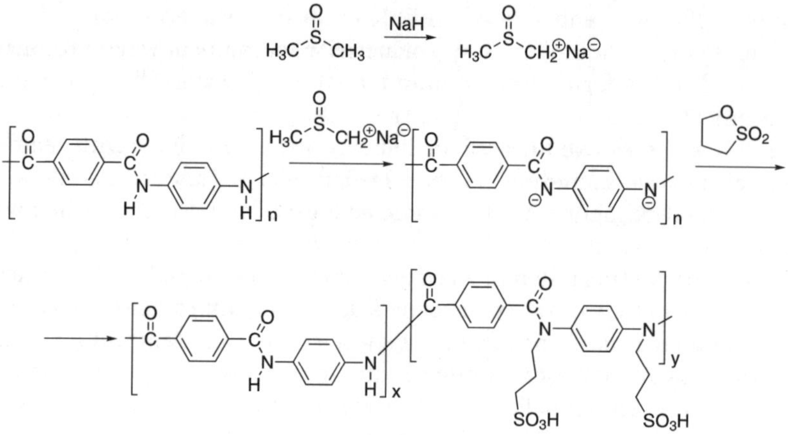

Scheme 20

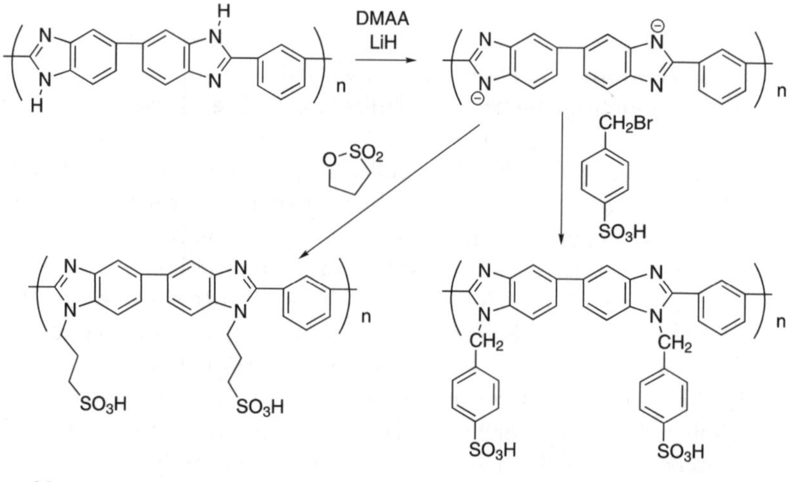

Scheme 21

The degree of alkylsulfonation as a function of the ratio of 1,3-propane sultone to PBI is represented in Fig. 18.

The degree of alkylsulfonation of NH groups in PBI was estimated considering the results of ^{1}H NMR study and elemental analysis. This parameter can be controlled easily by varying the ratio of 1,3-propane sultone to PBI. For instance, the alkylsulfonation level can be as high as 60 mol% at 1,3-propane sultone: PBI ratio of 5.0.

An attempt to synthesise ethylphosphorylated PBI using the above-mentioned treatment of PBI (Scheme 22) was reported [7].

The substitution reaction at the NH sites of benzimidazole rings was performed successfully, but the resulting polymer appeared to be insoluble in organic solvents. The reason for this can be aggregation of phosphoric acid

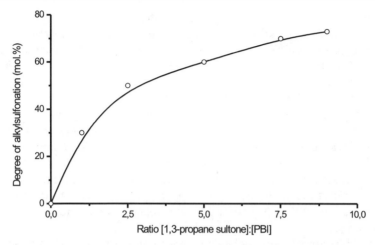

Fig. 18 Degree of alkylsulfonation of PBI as a function of 1,3-propane sultone: PBI ratio [7]

groups during the substitution reaction. Ethylphosphorylated PBI exhibited high proton conductivity (10^{-3} S cm^{-1}) even in the pellet form. According to the results obtained, the presence of polar phosphoric acid groups enhances the proton conductivity of polymer electrolytes.

Alkylsulfonation and arylsulfonation of the starting aromatic polymers was aimed at increasing their water uptake and proton conductivity while retaining high thermal stability. The polymers obtained were studied by TGA in inert and oxidative atmospheres [166]. Parent PBI exhibits very high thermal stability. In the inert atmosphere, the onset of its degradation occurs at 650 °C. The 5% mass loss is observed at 700 °C, and more than 80% of the polymer's initial mass is retained at 800 °C. Introduction of substituents that are not conjugated with the polymer backbones reduces the degradation temperatures in the inert medium, which is consistent with the expectations. The

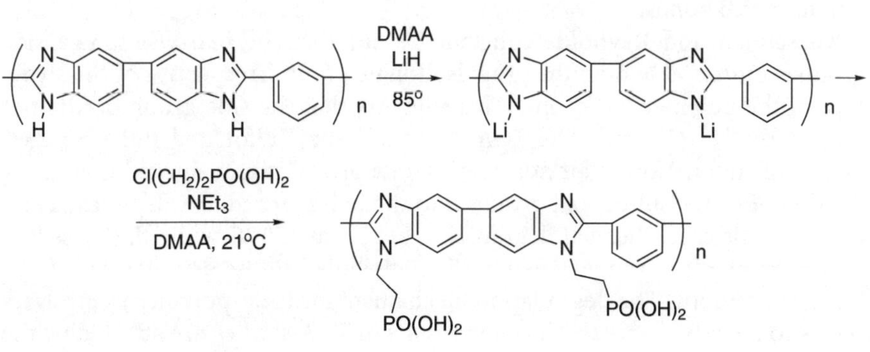

Scheme 22

degradation of poly[2,2'-*m*-phenylene-bi(N-benzylsulfonato)benzimidazolo-5,5'-diyl] with 22% substitution begins at 480 °C, while the onset of the mass loss process of poly{2,2'-*m*-phenylene-bi[N-(3-propylsulfo)benzimidazolo-5,5'-diyl]} with a substitution level of 54% is observed at 450 °C. After the removal of the substituting group the degradation slows down thus nearly 50–60% of the initial mass is retained at 800 °C.

The degradation of PBI in an oxidative atmosphere (dry air) begins at 520 °C, which is about 100 °C lower than the degradation temperature for this polymer in an inert medium. Degradation temperatures of substituted PBI in oxidative media are close to those of the unsubstituted ones. For all polymers, the mass loss in air is much higher than in nitrogen and the amount of residual char is much smaller. This happens primarily due to the lower stability of starting PBI in dry air and to some extent due to the introduction of substituents.

In an inert atmosphere, poly(*p*-phenylene terephthalamide) (PPTA) is stable below 550 °C. Rapid mass loss of the polymer (up to 50% of initial mass) begins at 600 °C. After modification with propylsulfonate side groups (66% substitution) the polymer is stable below 400 °C; only 40% of its initial mass is retained at 800 °C. The benzylsulfonated derivative of PPTA with a 66% substitution level is more thermally stable compared to the propylsulfonated derivative. Degradation of the latter begins at 470 °C. The decrease in mass of a sample down to 50% of its initial mass is observed at 800 °C. The degradation temperature of PPTA in a dry air atmosphere is 70 °C lower than in nitrogen [166]. Comparison of degradation processes of benzylsulfonated PPTA with 66% substitution in air and in nitrogen showed that the degradation in air begins at a lower temperature. The major difference is that the initial mass loss is higher, while the initial degradation is much smaller at high temperature, which is due to oxidative degradation of the polymer chains. Introduction of substituents into aromatic polymers reduces their thermal stability irrespective of the medium in which degradation occurs. This is the expected manner of changes in properties, since the side groups, especially sulfonic acid groups, are not stabilised by conjugation with the polymer backbones.

Gieselman and Reynolds concluded [166] that the benzylsulfonate side group is more stable than the propylsulfonate group irrespective of the structure of the polymer backbone. This suggests that the side group occurs not only at the N – C bond. The TGA study of benzylsulfonated PBI with 75% degree of sulfonation in air at a heating rate of 1 °C min^{-1} showed that introduction of benzylsulfonated groups into the polymer reduces its thermal stability. In this case, thermal degradation begins at 370 °C while the mass loss in the temperature range 370–420 °C is attributed to the degradation of sulfonic acid groups. The degradation mechanism for these polymer electrolytes seems to be very complex, since the results of TGA studies are affected by the residual water, impurities, sulfonation level and measurement conditions.

In air, arylsulfonated PBI is stable up to 350 °C, while benzylsulfonated PBI is stable up to 500 °C. These results are hard to compare because of different degrees of sulfonation of the PBI samples undergoing investigation. One can assume that benzylsulfonated PBI is less stable than propansulfonated PBI due to the presence of the weak Aryl–S bond. In fact, the degradation temperature of benzylsulfonated PBI is comparable with the degradation temperatures of polymeric electrolytes prepared via sulfonation with sulfuric acid.

The thermal stability of anhydrous propylsulfonated PBI (PBI-PS) in an atmosphere of nitrogen was studied by TGA at a heating rate of 5 °C min^{-1}. Prior to analysis, all samples were dried in vacuum at 60 °C for 48 h. However, this polymer is hydroscopic and it rapidly reabsorbs water after drying. Because of this, it was dried in situ and then differential thermal analysis was immediately performed.

In contrast to PBI, the degradation of PBI-PS was observed in the temperature range 400–450 °C. The decomposition temperature of PBI-PS decreases as the degree of alkylsulfonation increases to 400 °C (Fig. 19); however, it is higher than the degradation temperature of perfluorinated polymer electrolytes (nearly 280 °C).

Degradation of PBI-PS was studied by elemental analysis and FT-IR spectroscopy. It was found that the intensities of SO stretching vibrations decreased after heating the PBI-PS samples above 400 °C for 1 h. These results are similar to those reported by Gieselman and Reynolds [166] who found that the degradation of PBI-PS occurs due to desulfonation. Hence, alkylsulfonated PBI is more thermally stable than sulfonated aromatic polymer electrolytes characterised by a degradation temperature between

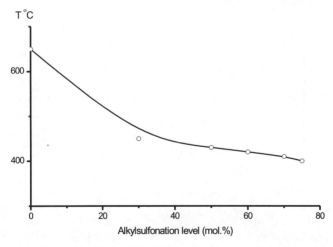

Fig. 19 Degree of alkylsulfonation of PBI-PS as a function of alkylsulfonation level [7]

200–350 °C. The thermal stability of alkylsulfonated polymer electrolytes can be attributed to the strong chemical bond between the alkyl and the sulfonic acid groups. The introduction of alkylsulfonic acid groups into thermostable polymers involving alkane sultone is one of the most important approaches to the preparation of thermostable proton-conducting polymer electrolytes.

Introduction of arylsulfonic and alkylsulfonic acid groups into aromatic polymer induces water absorption and makes them more hydroscopic. The water uptake of PBI-PS was determined by measuring the mass of the polymer before and after hydration. The dependence of the water uptake of PBI-PS on the relative humidity is presented in Fig. 20.

As can be seen, the water uptake changes with the relative humidity. The equilibrium water uptake of PBI-PS increases as the relative humidity and degree of alkylsulfonation increases. The water uptake of PBI-PS with an alkylsulfonation level of 73.1 mol % is 11.3 H_2O molecules per SO_3H group at room temperature and a relative humidity of 90% (cf. 11.0 molecules per SO_3H group for Nafion 117 membranes under the same conditions). This procedure was also employed for the synthesis of buthylsulfonated and (methyl)propylsulfonated PBI (PBI-BS and PBI-MPS, respectively) via butane sultone and methylpropane sultone. The water uptake of these polymers differ from that of PBI-PS and are 19.5 (PBI-BS) and 27.5 (PBI-MPS) H_2O molecules per SO_3H group at a relative humidity of 90%. The water uptakes of alkylsulfonated PBI depend on the length of alkyl chains and on the degree of chain branching, that is, as the chain length and the degree of alkyl chain branching increase, the water uptakes also increases. This is thought to be associated

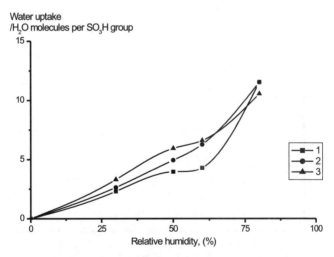

Fig. 20 Water uptake of PBI-PS as function of relative humidity at alkylsulfonation levels of 49.3 mol % (1), 61.5 mol % (2) and 73.1 mol % (3) [7]

with the greater flexibility of long alkyl chains and the larger amount of water absorbed in the cavities between the branched chains.

The specific role of the absorbed water in polymer electrolytes and the physical state of the water absorbed by polymer electrolytes were studied by IR [172] and ^{1}H NMR spectroscopy (low temperature relaxation time measurements) [173] and DSC [7]. The DSC curve of a hydrated PBI-PS film (73.1 mol %) containing 11.3 H_2O molecules per SO_3H group is shown in Fig. 21.

T_1 is the freezing temperature ($-36.6\,^{\circ}$C) and T_2 is the melting temperature ($-21.6\,^{\circ}$C).

The DSC curve of anhydrous PBI-PS exhibited no peaks, whereas the DSC curve of hydrated PBI-PS exhibited two peaks corresponding to phase transitions of absorbed water at -36.6 and $21.6\,^{\circ}$C that were attributed to the freezing and melting temperatures of the absorbed water, respectively.

A study of hydrated Nafion membranes under the same conditions revealed a phase transition at $0\,^{\circ}$C. These results indicate that the adsorbed water in the Nafion membranes is bound to a lesser extent compared to PBI-PS which can exist in the hydrated state even at elevated temperatures.

Wet PBI-PS films possess no electron conduction despite the fact that the main polymer chains are conjugated. To elucidate the nature of charge carriers in PBI-PS, the conductivity of PBI-PS films containing H_2O and D_2O was measured [7]. The results of measurements are presented in Fig. 22.

As can be seen, the conductivity of the films containing water increased with increasing water uptake and was higher than that of the PBI-PS films

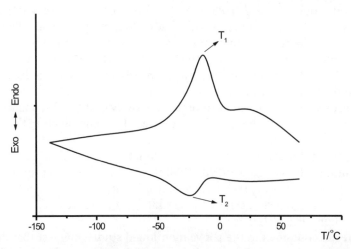

Fig. 21 DSC curve of hydrated PBI-PS (73.1%) film containing 11.3 H_2O molecules per sulfonic acid group [7]

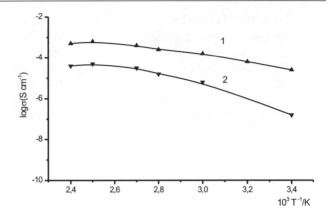

Fig. 22

containing D_2O in the same temperature range. This suggests that the charge carrier in hydrated PBI-PS is a proton (hydronium ion).

The temperature dependence of proton conductivity of PBI-PS containing the equilibrium amount of water is shown in Fig. 23.

Hydrated PBI-PS exhibits a high proton conductivity at room temperature. The conductivity of a PBI-PS sample containing 3.1 H_2O molecules per SO_3H group reached 10^{-5} S sm^{-1} at 80 °C and decreased slightly at higher temperatures due to a small loss of water ($\sim$ 10 mass %). The conductivity of a PBI-PS film containing more than 5.2 H_2O molecules per SO_3H group increased as the temperature increased and was as high as 10^{-3} S cm^{-1} at a temperature above 100 °C. The proton conductivity of a PBI-PS film containing 11.3 H_2O molecules per SO_3H group was 10^{-3} S cm^{-1}.

The water uptake of a PBI-PS film placed in an atmosphere with a relative humidity of 90% was compared with that of Nafion membranes. The proton conductivity of Nafion membranes was as high as 10^{-3} S cm^{-1} at temperature; however, it decreased due to the loss of absorbed water at temperatures above 100 °C. In contrast to this, hydrated PBI-PS exhibited a high proton conductivity at a temperature above 100 °C.

The large water uptake and proton conductivity of PBI-PS at a temperature above 100 °C are due to the specific properties of the polymer and the physical state of absorbed water.

The proton conductivity of benzylsulfonated PBI at different values of relative humidity has been studied [173]. It was found that the proton conductivity increases as the degree of the substitution increases. The polymer with a 75% substitution level exhibited a conductivity of 10^{-2} S cm^{-1} at 40 °C and a relative humidity of 100%.

The results obtained in the above mentioned studies suggest that the alkyl sulfonated aromatic polymer electrolyte exhibit sufficient thermal stabilities for fuel cell applications at 80 °C (a typical operating temperature for per-

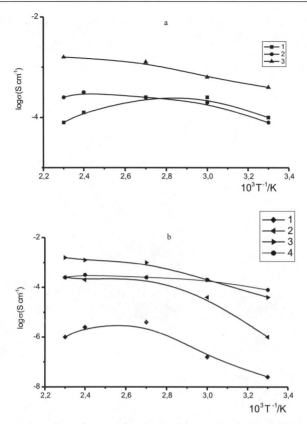

Fig. 23a,b Temperature dependences of proton conductivity of PBI-PS films with the same water uptake (48%) and different degrees of sulfonation (**a**) and with the same degree of sulfonation (73.1%) and different water uptakes (**b**) [7]; (**a**): degree of sulfonation (mol %): 49.3 (1), 61.5 (2), and 73.1 (3) (**b**): water uptake : 11.2 (1), 25.0 (2), 29.0 (3) and 48 (4)

fluorinated polymer electrolyte membranes). The water uptake and proton conductivity of this polymer are close to the corresponding values for perfluorinated polymer electrolytes at temperatures below 80 °C but are larger than the latter at temperatures above 80 °C.

The absorbed water molecules are more strongly bound to alkylsulfonated rather than perfluorinated polymers. One can assume that this is related to the difference in the absorbtion mechanisms and to the physical state of absorbed water in PBI-PS and perfluorinated polymer electrolytes.

A series of new sulfonated polymers where the sulfonic acid units are separated from the polymer main chains via short molecular spacers were developed [174, 175] using treatment of polyethersulfones with BuLi leading to the formation of lithiated polymers and subsequent transformations under the action of bromoalkanesulfonates, butansultone or 2-sulfobenzoic acid cyclic anhydride.

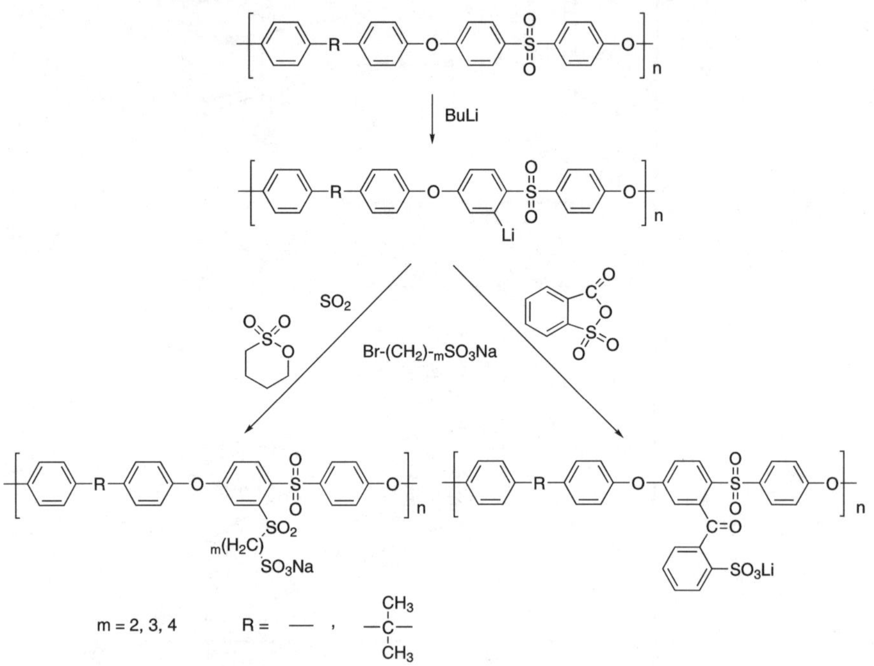

Scheme 23

Recently conjugated polyelectrolytes containing alkylsulphonic and alkyl-phosphonic groups have received considerable interest [176–180]. Sulfonate-substituted poly(*p*-phenylene) was prepared [176] using Suzuki co-polymerisation of 1,4-benzenediboronic acid with sulfonate monomer in three steps starting from 1,4-dimethoxybenzene:

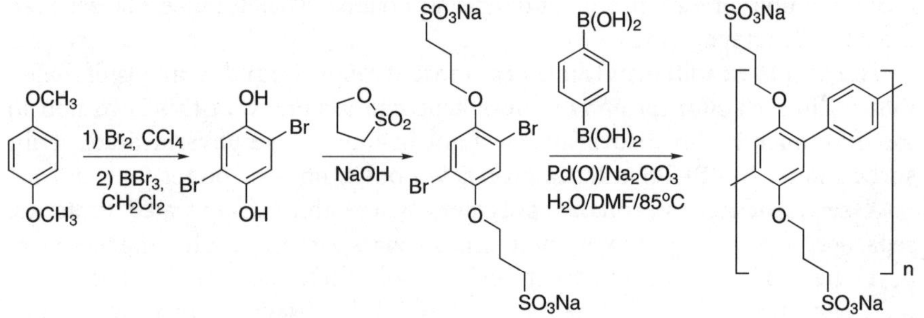

Scheme 24

Several years later, Reynolds and co-workers [179] extended this work to prepare another sulfonated poly(p-phenylene) using the same approach by replacing 1,4-diboronic acid with 4,4′-biphenyl diboronic ester:

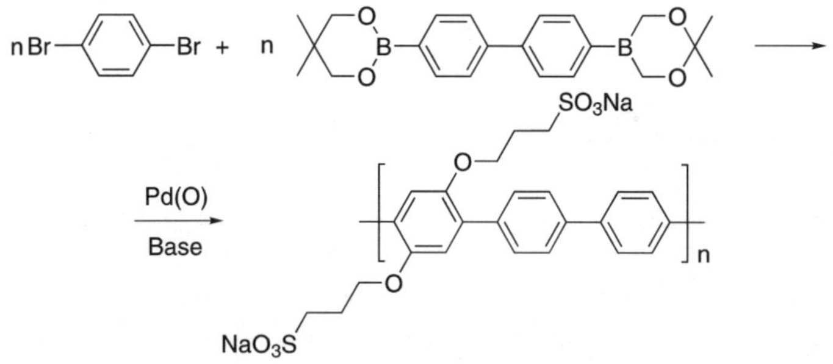

Scheme 25

In a recent paper Shanze and co-workers [177] reported the synthesis of poly(p-phenylene ethynylene) which was obtained by Sonogashira coupling in accordance with the following scheme:

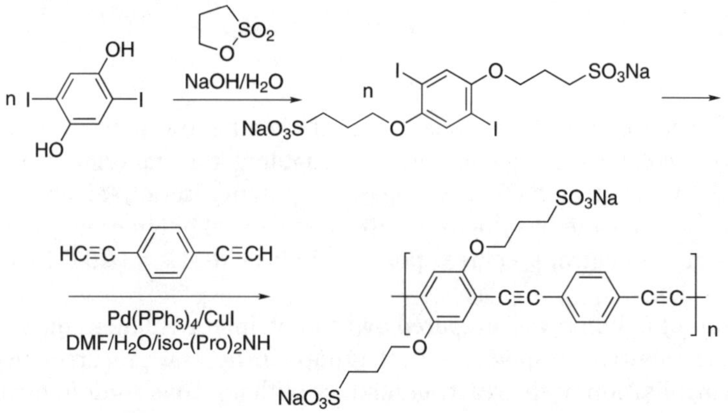

Scheme 26

The resulting polymer was obtained in a 68% yield, and it was soluble in water and low molecular weight alcohols. A molecular weight of 100 kD was estimated by the polymer's ultrafiltration properties and by iodine end-group analysis.

In a recent preliminary report [178] Shanze et al. described poly(phenylene ethynylene) which features phosphonate groups appended to the polymer backbone:

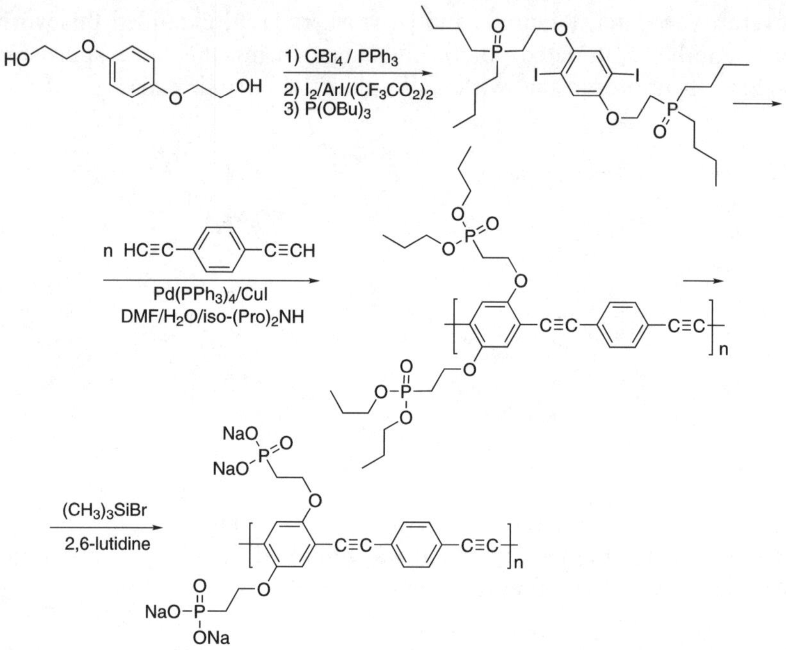

Scheme 27

The phosphonate polymer was prepared via a neutral precursor polymer, which was soluble in organic solvents, enabling the material to be characterised by NMR and GPC. Sonogashira polymerisation of phosphonate monomer and 1,4-diethynylbenzene afforded neutral polymer in a 46% yield. Analysis of the neutral precursor polymer indicated Mw = 18.3 Kd and polydispersity 2,9.

The target polymer was prepared by bromotrimethylsilane-induced cleavage of the *n*-butyl phosphonate ester groups in neutral precursor polymer. After neutralisation of the reaction mixture with aqueous sodium hydroxide, the target polymer has exhibited good solubility in water.

4
Proton-Exchanging Electrolyte Membranes Based On Polymer Complexes

Proton-conducting membranes used in PEMFC operate under severe conditions (see above). Recently, complexes of basic polymers with strong acids have attracted considerable interest. Such complexes are characterised by stable electrochemical properties and large water uptakes at high temperature.

Recently, new proton-conducting polymer electrolyte membranes based on PBI – orthophosphoric and other strong acid complexes have been proposed for use in PEMFCs [181–188].

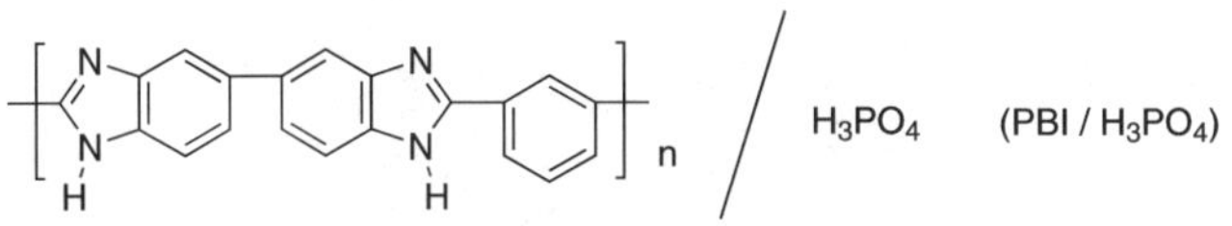

Fig. 24

The most important advantages of this polymer electrolyte over perfluorinated polymer electrolytes and other acid – basic polymer complexes are that PBI/H_3PO_4 possesses conductivity even at low activity at water and high thermal stability of these systems. The materials based on these complexes are expected to operate over a wide range from room to high temperature in both humid and dry gas. Such complexes are prepared by immersing PBI films into phosphoric acid solutions. In particular, the preparation of PBI-strong acid complexes by immersion PBI films into solutions of strong acids in methanol was reported [187, 189]. The absorption level of strong acid molecules increased with an increase in the concentration of the strong acid and reached up to 2.9 molecules per repeating unit for polymer complexes PBI/H_3PO_4. IR spectroscopy study of the complexes revealed that the acid molecules, except for H_3PO_4 protonate the nitrogen atoms in the imidazole ring. Phosphoric acid (H_3PO_4) is incapable of protonating the imidazole groups in PBI but interacts with them via the formation of strong hydrogen bonds between NH and OH groups.

PBI films doped with phosphoric acid were prepared by immersion of PBI films in aqueous solutions of phosphoric acid for at least 16 h [181–185]. Upon equilibration in a 11 M H_3PO_4 solution a doping level of $\sim$ 5 phosphoric acid molecules per repeating unit of the polymer was achieved.

PBI membranes loaded with high levels of phosphoric acid were prepared using a new sol-gel process [190]. This process, termed the PPA process, uses PPA as the condensing agent for the polycyclocondensation and the membrane casting solvent. After casting, absorption of water from the atmosphere causes hydrolysis of the PPA to phosphoric acid.

The thermal stability of PBI-strong acid polymer complexes was studied by TGA and DTA. Fig. 25 presents the TGA curves of polybenzimidazole and its complexes with strong acids.

As can be seen, PBI exhibits an extremely high thermal stability over the entire temperature range. Small mass losses by all samples at temperatures below 200 °C are due to the loss of water and solvent present in the membranes.

Typical proton-conducting polymer electrolytes undergo considerable degradation in the temperature range under study.

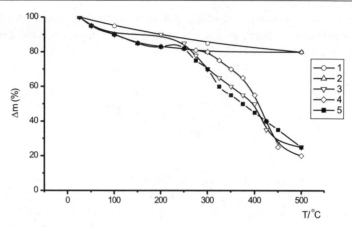

Fig. 25 TGA curves of PBI (1) and its complexes with H_3PO_4 (2), H_2SO_4 (3), $MeSO_3H$ (4) and $EtSO_3H$ (5) [7].

A decrease in the degradation temperature of polymer complexes PBI/ H_3PO_4 was expected because of the complexation of acid molecules which easily corrodize and oxidise the polymer macromolecules. However, no degradation was observed under the nitrogen atmosphere. At the same time, thermal decomposition of PBI complexes with H_2SO_4, $MeSO_3H$, $EtSO_3H$ begins at 330, 240 and 220 °C respectively. After thermal decomposition of these polymer complexes in the temperature range 220–400 °C the residues were 50% of the initial masses of the samples. Therefore, complexation of PBI with H_2SO_4, $MeSO_3H$, $EtSO_3H$ results in a loss of thermal stability. The decomposition of complexes is first of all due to elimination of acid molecules. This assumption was confirmed by the results of elemental analysis. At temperatures above 400 °C, the PBI chains gradually decompose under the action of high temperature and strong acids.

Complexes PBI/H_3PO_4 are thermally stable up to 500 °C. It was found that treatment of PBI with a phosphoric acid solution (27 mass %) improved the thermal stability of the polymer [191]. This was associated with the formation of benzimidazonium cations. Samms et al. [185] studied the thermal stability of polymer complexes and showed that these complexes are promising for use as polymer electrolytes in the hydrogen-air and methanol fuel cells. To simulate the operating conditions in a high-temperature PEMFC, the polymer complexes PBI/H_3PO_4 were coated with platinum black, doped with phosphoric acid (4.8 H_3PO_4 molecules per repeating unit of PBI) and heated in an atmosphere of nitrogen and 5% hydrogen or in air in the TGA analyser. The degradation products were identified by mass spectrometry. In all cases the mass loss below 400 °C was found to be due to the loss of water. In addition, it was found that polymer complexes PBI/H_3PO_4 coated with platinum black are thermally stable up to 600 °C.

Variation of the conductivity of polymer complexes PBI/H_3PO_4 as a function of water vapour activity, temperature and acid doping level was studied [183]. It was shown that the conductivity of heavily-doped complexes (500 mol %) is nearly twice as high as that of the film doped to 338 mol % at the same temperature and humidity. For instance, the conductivity of PBI doped with 500 mol % H_3PO_4 (5H_3PO_4 molecules per repeating unit of PBI) is 3.5×10^{-2} S cm^{-1} at 190 °C and water vapour activity of 0.1.

Raising the temperature and water vapour activity causes an increase in the conductivity of the polymers irrespective of the doping level of PBI with phosphoric acid. In addition, it was found that crossover of methanol molecules through the polymer complexes (Fig. 24) is by an order of mag-

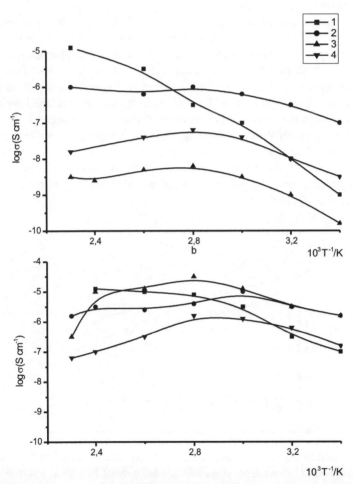

Fig. 26a,b Temperature dependence of proton conductivity of anhydrous (**a**) and hydrated (**b**) PBI complexes with H_3PO_4 (1), H_2SO_4 (2), $EtSO_3H$ (3), $MeSO_3H$ (4) [7]

nitude smaller than in the case of perfluorinated polymer electrolytes and that the mechanical strength of such complexes is three orders of magnitude higher compared to that of Nafion membranes.

The proton conductivity of PBI polymer complexes prepared by the interaction of PBI with methanol solutions of strong acids was studied [7]. The temperature dependences of the conductivities of anhydrous PBI-strong acid polymer complexes are shown in Fig. 26a.

All anhydrous polymer complexes of PBI with strong acids possess a proton conductivity of the order of 10^{-6}–10^{-9} S cm^{-1} at 100 °C. The conductivity of polymer complexes PBI/H$_3$PO$_4$ can be as high as 10^{-5} S cm^{-1} at 160 °C, whereas other PBI-acid complexes showed a decrease in the conductivity at temperature above 80 °C. These results point to high thermal stability of polymer complexes PBI/H$_3$PO$_4$.

To prepare hydrated systems, the films of PBI-strong acid polymer complexes were placed in a desiccator with a relative humidity of 90% for 72 h. The water uptake of the complexes were 13–26 mass %. The proton conductivity of the hydrated PBI-strong acid polymer complexes was found to be nearly an order of magnitude higher than the conductivity of anhydrous polymer complexes (Fig. 26b). This difference can be explained by the improvement of charge carrier generation in the absorbed water.

Changes in the proton conductivity at room temperature are especially remarkable. Fig. 27 presents the temperature dependences of the conductivities of anhydrous complexes PBI/H$_3$PO$_4$ with different acid contents. As can be seen, the conductivity of polymer complexes PBI/H$_3$PO$_4$ increases with the concentration of H$_3$PO$_4$.

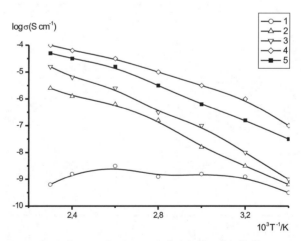

Fig. 27 Temperature dependence of proton conductivity of anhydrous PBI/H$_3$PO$_4$ complexes containing 1.4 (1), 2.0 (2), 2.7 (3), 2.3 (4), 2.9 (5) H$_3$PO$_4$ molecules per PBI units [7]

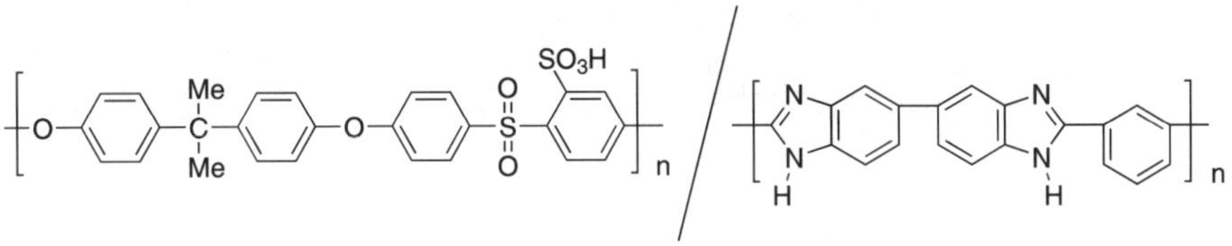

Fig. 28

The temperature dependences of the conductivities of polymer complexes PBI/H_3PO_4 are quite different: in this temperature range the conductivity is low. This suggests that two H_3PO_4 molecules quantitatively react with the PBI units containing two imidazole groups. As a consequence, an excess of H_3PO_4 determines the necessary proton conductivity. A study of PBI/H_3PO_4 polymer complexes by FT-IR spectroscopy showed that the spectra exhibited three characteristic absorption maxima near $1090\,cm^{-1}$ HPO_4^{-2}, $1008\,cm^{-1}$ (P-OH) and $970\,cm^{-1}$ $H_2PO_4^{-1}$ [192–195]. As the concentration of H_3PO_4, the intensity of the absorption maxima of HPO_4^{-2} and $H_2PO_4^-$ increases. This suggests that proton conductivity can occur by the Grotthus mechanism [196] involving an exchange of protons between H_3PO_4 and PO_4^{-2} or $H_2PO_4^-$.

Anhydrous sulfonated aromatic polymers are highly brittle. Recently [197], new materials with high mechanical strength were reported. They were prepared using a polymer blending technique by combining PBI and sulfonated polymers (S-PEEK or *ortho*-sulfonated polysulfone) (Fig. 28).

Such polymer blends exhibit high proton conductivities, moderate swelling values and high thermal stabilities. The specific interaction of SO_3H groups with basic nitrogen atoms was confirmed by FT-IR spectroscopy. The acid-base interaction between the sulfonated polymer and PBI provided a material with high mechanical strength and thermal stability.

Along with PBIs for the preparation of basic polymer strong acid complexes polyphenylquinoxaline [198],

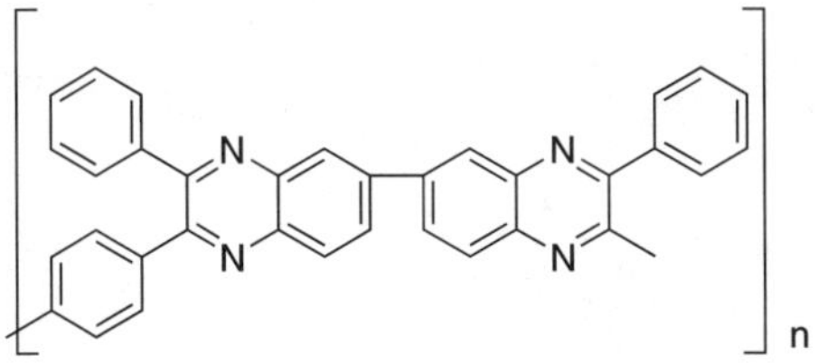

Fig. 29

poly-1,3,4-oxadiazole [199]

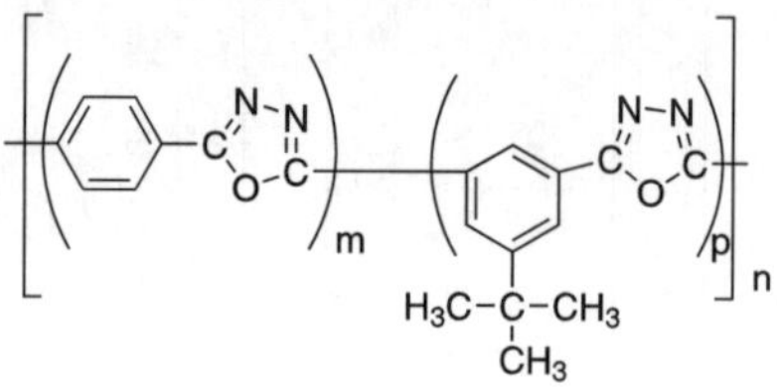

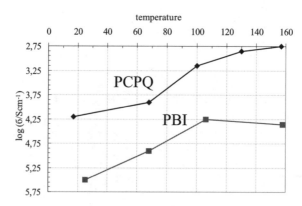

Fig. 30

and pyrrole-containing polyquinoline [200] were used:

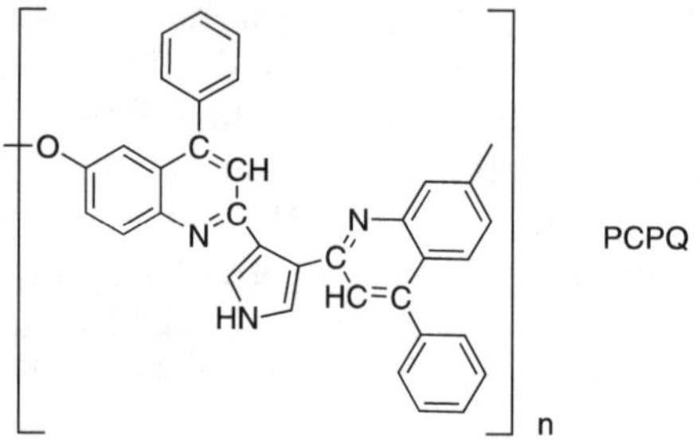

PCPQ

Fig. 31 Temperature dependence of conductivity for PCPQ/H_3PO_4 and PBI/H_3PO_4 complexes at 3.8 and 1.5 mol unit^{-1}

The last polymer with good solubility in DMAC was chosen to prepare a membrane and to examine the proton conductivity. For comparison, the PBI was also measured under the same experimental conditions.

Temperature dependences of conductivity for PCPQ/H_3PO_4 and PBI/H_3PO_4 complexes at 3.8 and 1.5 mol unit^{-1}, respectively, are shown in Fig. 32:

Fig. 32

As can be seen from Fig. 32, the conductivities of PCPQ/H_3PO_4 and PBI/H_3PO_4 complexes increased with temperature, and the PCPQ/H_3PO_4 complex exhibited a higher conductivity compared with that of the PBI/H_3PO_4 complex, reaching 1.5×10^{-3} S cm^{-1} at 157 °C, while the conductivity of PBI/H_3PO_4 complex was 7×10^{-5} S cm^{-1} at 150 °C. In addition, under the same experimental conditions PCPQ could complex more H_3PO_4 (3.8 mol unit^{-1} than PBI (1.5 mol unit^{-1}. That may be the main reason for PCPQ/H_3PO_4 having a higher conductivity.

5
Fuel Cell Applications of Proton-Exchanging Membranes Based On Aromatic Condensation Polymers

Two blend polymer electrolytes containing acid and basic functional groups (90 mass % PEEK and 10 mass % PBI or 95 mass % PES and 5 mass % PBI) were applied in H_2/O_2 fuel cells. The current vs. voltage curves of the membranes in the fuel cells were comparable with that of Nafion 112 membranes [197].

Fuel cell tests of membranes based on sulfonated PES showed [7] a cell voltage of 550 mV at a current density of 700 mA cm^{-2} (atmospheric pressure, humidified gases, 70 °C). No significant loss of membrane performance was observed after long-term operation (1000 h) under fuel cell conditions.

The maximum power of fuel cells with S-PPBP membranes reaches 0.3 W cm^{-2} at a current density of 800 mA cm^{-2}. The conductivity of the electrolyte membranes was 3×10^{-3} S cm^{-1}; the membrane thickness and surface area were 0.01 cm and 3.15 cm^2, respectively.

The maximum power of fuel cells H_2/O_2 and CH_3OH/O_2 with a membrane based on polymer complexes PBI/H_3PO_4 [7] was as high as 0.25 W cm^{-2} at a current density of 700 mA cm^{-2}. The electrical resistance of electrolyte membranes was 0.4 Ω, the thickness and surface area of the membranes were 0.01 cm and 1 cm^2, and the doping level was 500 mol %. The measured electrical resistance of the cell was equivalent to a conductivity of 0.025 S cm^{-1}. It was found that the electrical resistance of the fuel cell is independent of the water content in the gas (water produced at the cathode is sufficient for maintaining the necessary conductivity of the electrolyte). This type of fuel cell was characterised by continuous operation at a current density of 200 mA cm^{-2} over a period of 200 h (and for longer time periods) without reduction of the membrane performance.

The power of CH_3OH/O_2 fuel cells at 200 °C and atmospheric pressure reached 0.1 W cm^{-2} at a current density of 250–500 mA cm^{-2}. The conductivity of membranes operating under these conditions remained constant in the temperature range 30–140 °C.

6
Conclusions

To summarise, the aforesaid shows that aromatic condensation polymers can be thought of as candidates for fuel cell applications; the structure of ACP-based polymer electrolytes can be modified with ease; these polymer electrolytes possess a large water uptake and high proton conductivity at high temperature and low humidities, as well as sufficient thermal and chemical stability. Further investigations to design proton-conducting materials exhibiting long-term thermal stability and mechanical strength, capable of operating at high temperatures without humidification and further improvement of fuel cells are required.

References

1. Kerres JA (2001) J Membrane Sci 185:3
2. Riedinger H, Faul W (1988) J Membrane Sci 36:5
3. Nolte R, Ledjeff K, Bauer M, Mülhaupt R (1993) J Membrane Sci 83:211
4. Prater KB (1994) J Power Sources 51:129
5. Watkins S (1993) In: Blumen LG, Mugerwa MN (eds) Fuel cell systems. Plenum, New York, p 493
6. Lassegues JC (1992) In: Colombon P (ed) Proton Conductors: Solids, Membranes and Gels. Cambridge University Press, Cambridge, p 311
7. Rikukawa M, Sanui K (2000) Prog Polym Sci 25:1463
8. Savadogo O (1998) J New Mater Electrochem Syst 1:66
9. Higuchi M, Minoura N, Kinoshita T (1994) Chem Lett 2:227
10. Zawodzinski TA, Derouin C, Radzinski S, Sherman RJ, Smith UT, Springer TE, Gottesfeld S (1993) J Electrochem Soc 140:1041
11. Yamabe M, Migake H (1994) In: Bomks RE, Smart BE, Tatlow JC (eds) Organofluorine Chemistry. Principles and Commercial Applications. Plenum, New York, p 403
12. Kirsh YuE, Smirnov SA, Popkov YuM, Timashev SF (1990) Russ Chem Rev 59:970
13. Grot WG (1994) Macromol Symp 82:161
14. Steck A (1995) In: Savadogo O, Roberge PR, Vezirogly TN (eds) Proceedings of the 1st International Symposium on New Materials for Fuel Cell Systems, Membrane Materials in Fuel Cells. Montreal, Canada, p 74
15. Eisenberg A, Yeagger HL (1982) Perfluorinated Ionomer Membranes, ACS Symposium Series Vol 180. Am Chem Soc, Washington DC
16. Shoesmith JP, Collins RD, Oakley MJ, Stevenson DK (1994) J Power Source 49:129
17. Hay AS (1967) Adv Polymer Sci 4:496
18. Maiti S, Mandal B (1986) Prog Polym Sci 12:111
19. Lee H, Stoffey D, Neville K (1967) New Linear Polymers. McGraw-Hill, New York
20. Frazer AH (1968) High Temperature Resistant Polymers. Wiley Interscience, New York
21. Korshak VV (1969) Termostoikie Polymery (Thermally Stable Polymers). Khimiya, Moscow
22. Cassidy PE (1980) Thermally Stable Polymers. Marcel Dekker, New York

23. Gritchley JP, Wright WW (1983) Heat-Resistant Polymers. Plenum, New York
24. Bühler K-U (1978) Spezialplaste. Akademie-Verlag, Berlin
25. Rusanov AL, Tugushi DS, Korshak VV (1988) Uspekhi Khimii Poligeteroarilenov (Progress in the Polyheteroarylenes Chemistry). Tbilisi State University, Tbilisi
26. Korshak VV, Rusanov AL (1984) Vysokomol Soedin. Ser A 26:3
27. Rusanov AL (1986) Vysokomol Soedin. Ser A 28:1571
28. Schluter AD (2001) J Polym Sci Polym Chem 39:1533
29. Lu F (1998) J Macromol Sci Rev Macromol Chem Phys C38(2):143
30. Luise RR (1996) Applications of High Temperature Polymers. CRC Press, Boca Raton, FL, p 272
31. Kreuer KD (2001) J Membrane Sci 185:29
32. Kreuer KD (1996) Chem Mater 8:610
33. Rusanov AL, Likhatchev DYu, Müllen K (2002) Russ Chem Revs 71:761
34. Qi Z, Pickup PG (1998) Chem Comm 1:15
35. Kobayashi T, Rikukawa T, Sanui K, Ogata N (1998) Solid State Ionics 106:219
36. Sochilin VA, Pebalk AV, Semenov VI, Kardash IE (1991) Vysokomol Soedin, Ser A 33:1536
37. Sochilin VA, Pebalk AV, Semenov VI, Sevast'yanov MA, Kardash IE (1993) Vysokomol Soedin. Ser A 35:1480
38. US Patent 3 259 592 (1966) Chem Abstr 65:13902a (1966)
39. US Patent 3 709 841 (1970) Chem Abstr 74:43215u (1971)
40. US Patent 3 780 496 (1973) Chem Abstr 80:72484r (1974)
41. Chludzinski PJ, Ficket AF, La Conti AB (1971) Am Chem Soc Polymer Prepr 12:276
42. Chalk AJ, Hay AS (1982) J Polym Sci, Ser A 7:5843
43. Myataki K, Oyaizu K, Tsuchida E, Hay AS (2001) Macromolecules 34:2065
44. Shaikh AG, Hay AS (2002) J Polym Sci Polym Chem 40:496
45. Wang L, Meng YZ, Wang SJ, Shang XY, Hay AS (2004) Macromolecules 37:3151
46. Bredas IL, Chomce RR, Silbey R (1982) Phys Rev B 26:5843
47. Kobayashi H, Tomita H, Moriyama H, Kobayashi A, Watanabe T (1994) J Am Chem Soc 116:3153
48. Wong F, Roovers J (1993) Macromolecules 26:5295
49. Bailly C, Williams DJ, Karasz FF, MackNight WJ (1987) Polymer 28:1009
50. Jin X, Bishop MT, Ellis TS, Karasz FE (1985) Br Polymer J 17:4
51. Lee J, Marvel CS (1983) J Polymer Sci, Polym Chem Ed 21:2189
52. Litter MI, Marvel CS (1985) J Polymer Sci, Polym Chem Ed 23:2205
53. Devaux J, Delimoy D, Daoust D, Legras R, Mercier JP, Strazielle C, Neild E (1985) Polymer 26:322
54. Zaidi SMJ (2003) Arab J Sci Eng B (Eng) 28(2B):183
55. Bishop MT, Karasz FE, Russo PS, Langley KH (1985) Macromolecules 18:86
56. Shibuya N, Porter RS (1992) Macromolecules 25:6495
57. Jia LXuX, Zhang HXuJ (1996) J Appl Polym Sci 60:1231
58. Trotta F, Drudi EG, Moraglio E, Baima P (1998) J Appl Polym Sci 62:70
59. Sakaguchi Y, Kitamura R, Takase S (2003) Am Chem Soc Polymer Prepr 44(2):785
60. US Patent 4 273 903 (1980) Chem Abstr 93:24029j (1980)
61. US Patent 4 625 000 (1987) Chem Abstr 106 85 292
62. US Patent 4 413 106 (1983) Chem Abstr 100 35 025
63. Noshay A, Robeson LM (1976) J Appl Polym Sci 20:1885
64. Johnson BC, Tram C, Yilgor J, Iqbal M, Wightman JP, Lloyd DR, McGrath JE (1983) Am Chem Soc Polymer Prepr 24(1):31

65. Johnson BC, Yilgor J, Tram C, Iqbal M, Wightman JP, Lloyd DR, McGrath JE (1984) J Polym Sci Polym, Chem Ed 22:721
66. Mottet C, Revillon A, Perchec P, Lauro ME, Guyot A (1982) Polym Bull 8:511
67. Arnold Jr C, Assinic RA (1988) J Membrane Sci 38:71
68. Zschocke P, Guellmalz D (1985) J Membrane Sci 22:325
69. Genova-Dimitrova P, Baradie B, Foscallo D, Poisington C, Sanchez JY (1976) J Membrane Sci 185:59
70. Poppe D, Frey H, Kreuer KD, Heinzel A, Mülhaupt R (2002) Macromolecules 35:7936
71. Harrison WL, Wong F, O'Connor R, Arnett NY, Kim YS, McGrath JE (2003) Am Chem Soc Polymer Prepr 44(1):849
72. Harrison WL, O'Connor R, Arnett NY, McGrath JE (2002) Am Chem Soc Polymer Prepr 43(2):1159
73. Ghassemi H, Ndip G, McGrath JE (2003) Am Chem Soc Polymer Prepr 44(1):814
74. Sakaguchi Y, Kitamura K, Takase S (2003) Am Chem Soc Polymer Prepr 44(2):783
75. Qi Z, Lefebre MC, Pickup PG (1998) J Electroanalytical Chem 459:9
76. Russian Patent 1 819 418 (1992) Byull Izobret (14):61
77. Belomoina NM, Rusanov AL, Yanul' NA, Kirsh YuE (1996) Vysokomol Soedin Ser B 38:355
78. Kirsh YuE, Yanul' NA, Belomoina NM Rusanov AL (1996) Elektrokhimiya 32:169
79. Kopitzke RW, Lincous CA, Anderson H, Randolph N, Gordon L (2000) J Electrochem Soc 147(5):1677
80. US Patent 4 634 530 (1987) Chem Abstr 106 157 701
81. Quante H, Schlichting P, Rohr U, Geerts Y, Müllen K (1996) Macromol Chem Phys 197:4029
82. Wei XL, Wamg YZ, Long SM, Bobeczko C, Epstein AJ (1996) J Am Chem Soc 118:2545
83. Gilbert EE (1965) Sulfonation and Related Reactions. Wiley-Interscience, New York
84. Cerfontain H (1968) Mechanistic Aspects in Aromatic Sulfonation and Desulfonation. Wiley-Interscience, New York
85. Taylor R (1972) In: Bamford CH, Tipper CFH (ed) Chemical Kinetics. Reactions of Aromatic Compounds. v 13, p 56.
86. Lee J, Marvel CS (1984) J Polym Sci, Polym Chem Ed 22:295
87. Johnson BC, Yilgor I, Iqbal M, Wrightman JP, Lloyd D, McGrath JE (1984) J Polym Sci Polym Chem Ed 22:72
88. Bially C, Williams D, Karasz FE, MacKnight WJ (1987) Polymer 28:1009
89. Thaler WA (1982) J Polym Sci 20:875
90. Thaler WA (1983) Macromolecules 16:623
91. Bishop MT, Karasz FE, Russo PS, Langley KH (1985) Macromolecules 18:86
92. Devaux J, Delimoy D, Daoust D, Legras R, Mercier JP, Strazielle C, Neild E (1985) Polymer 26:1994
93. Ogava T, Marvel CS (1985) J Polym Sci Polym Chem Ed 23:1231
94. Shibuya M, Porter RS (1992) Macromolecules 25:6495
95. Miyatake K, Iyotani H, Yamamoto K, Tsuchida E (1996) Macromolecules 29:6969
96. Miyatake K, Shouji E, Yamamoto K, Tsuchida E (1997) Macromolecules 30:2941
97. Japan Patent 11-116 679 (1999) Chem Abstr 130 325 763
98. Japan Patent 11-067 224 (1999) Chem Abstr 130 239 961

99. Rusanov AL, Valetskiy PM, Belomoina NM, Keshtov ML, Yanul NA, Likhatchev DYu (2003) Proceedings Conf. Advances in Materials for Proton Exchange Membrane Fuel Cell Systems Asilomar, CA, USA, 23–26 February 2003, Prepr 15
100. Rusanov AL, Keshtov ML, Belomoina NM, Mikitaev AK (1997) Polymer Science A 39:1046
101. Rusanov AL, Keshtov ML, Belomoina NM, Petrovskiy PV (1999) Polymer Science A 41:61
102. Rusanov AL, Keshtov ML, Belomoina NM, Likhatchev DYu (2004) Abstracts Polycondensation Conference, Roanoke VA, USA, 26–29 September 2004
103. Rulkens R, Schulze M, Wegner G (1994) Macromol Rapid Commun 15:669
104. Rulkens R, Wegner G, Enkelmann V, Schulze M (1996) Ber Bunsenges Phys Chem 100:707
105. Bockstaller M, Köhler W, Wegner G, Fytas G (2000) Macromolecules 34:6359
106. Venkatasubramanion N, Dean DR, Arnold FE (1996) Am Chem Soc Polym Prepr 37(1):354
107. Venkatasubramanian N, Dean DR, Price GE, Arnold FE (1997) High Perform Polym 9:291
108. Guyu Xiao, Guomin Sun, Deye Yan (2002) Macromol Rapid Commun 23:488
109. Gao Y, Robertson GP, Guiver MD, Jian X, Mikhailenko SD, Wang K, Kaliagin S (2003) J Polym Sci Polym Chem Ed 41:2731
110. Ueda M, Toyota H, Ochi T, Sugiyama J, Yonetake K, Masuko T, Teramoto T (1993) J Polym Sci, Polym Chem Ed 31:853
111. McGrath JE, Formato R, Kovar R, Harrison W, Mechant JB (2000) Am Chem Soc Polym Prepr 41(1):237
112. Wang F, Glass T, Li X, Hickner M, Kim YS, McGrath JE (2002) Am Chem Soc Polym Prepr 43(1):492
113. Harrison W, Wang W, Kim YS, Hickner M, McGrath JE (2002) Am Chem Soc Polym Prepr 43(1):700
114. Harrison WL, Summer MJ, Hill M, Kim YS, Hickner M, Tchathova CN, Dong L, Riffle JS, McGrath JE (2003) Am Chem Soc Polym Prepr 42(2):647
115. Mecham J, Shobha HK, Wong F, Harrison W, McGrath JE (2000) Am Chem Soc Polym Prepr 41(2):1388
116. Wong F, Mecham J, Harrison W, McGrath JE (2000) Am Chem Soc Polym Prepr 41(2):1401
117. Guyu Xiao, Guomin Sun, Deye Yan, Pinfong Zhu, Ping Tao (2002) Polymer 43:5335
118. Vandenberg EJ, Dively WR, Filar LJ, Patel SR, Barth HB (1987) Polym Mater Sci Eng 57:139
119. Salamone JC, Krause SF, Richard RE, Clough SB, Waterson AL, Vandenberg EJ, Dively WR, Filar LJ (1987) Polym Mater Sci Eng 57:144
120. Salamone JC, Li CK, Clough SB, Bennet SL, Waterson AL (1988) Am Chem Soc Polym Prepr 29(1):273
121. Kirsh YuE, Fedotov YuA, Yudina NA, Artemov DYu, Yanul NA, TN Nekrasova (1991) Vysokomol Soedin Ser. A 38:1127
122. Sarkar N, Kershner LD (1996) J Appl Polym Sci 62:393
123. Shobha NK, Somkarapandion M, Glass TE, McGrath JE (2001) Am Chem Soc Polym Prepr 41(2):1298
124. Faure S, Mercier R, Pineri M, Sillion B (1996) In: The 4th Europian Technical Symposium on Polyimides and Other High Performance Polymers. Montpellier, France, p 414

125. Genies C, Mercier R, Sillion B, Petioud R, Cornet N, Gebel G, Pineri M (2001) Polymer 42:5097
126. Gunduz N, McGrath JE (2000) Am Chem Soc Polym Prepr 41(2):1565
127. Salle R, Sillion B (1974) French Patent 2 212 356
128. Faure S, Mercier R, Albert P, Pineri M, Sillion B (1996) French Patent 9 605 707
129. Faure S, Cornet N, Gebel G, Mercier R, Pineri M, Sillion B (1997) In: Proceedings of Second International Symposium on New Materials for Fuel Cell and Modern Battery System. Montreal, Canada, p 818
130. Vallejo E, Pourcelly G, Gavach C, Mercier R, Pineri M (1999) J Membr Sci 160:127
131. Genies C, Mercier R, Sillion B, Cornet N, Gebel G, Pineri M (2001) Polymer 42:359
132. Timofeeva GI, Ponomarev II, Khokhlov AR, Mercier R, Sillion B (1996) Macromol Symp 106:345
133. Zhong Y, Litt M, Jiong J, Savinell RF, Wainright JS (1999) In: Proceedings of 5th European Technical Symposium on Polyimides and Other High Performance Polymers. Montpellier, France, p 268
134. Zhong Y, Litt M, Savinell RF, Wainright JS, Vendramin J (2000) Am Chem Soc Polym Prepr 41(2):1561
135. Kim HJ, Litt M (2001) Am Chem Soc Polym Prepr 42(2):486
136. Shobha HK, Sacarapandian M, Glass TE, McGrath JE (2000) Am Chem Soc Polym Prepr 41(2):1298
137. Hong Y-T, Einsla B, Kim Y, McGrath JE (2002) Am Chem Soc Polym Prepr 43(1):666
138. Solomin VA, Lyakh EN, Zhubanov BA (1992) Polymer Sci A 34:274
139. Guo X, Fang J, Watari T, Tanaka K, Kita H, Okamoto K-I (2002) Macromolecules 35:6707
140. Fang J, Guo X, Watari T, Tanaka K, Kita H, Okamoto K (2002) Macromolecules 35:9022
141. Yin Y, Fang J, Tanaka K, Kita H, Okamoto K (2002) Polymer Prep Japan 51:2782
142. Fang J, Guo X, Watari T, Tanaka K, Kita H Okamoto K (2003) In: Mittal KL (ed) Polyimides and Other High Temperature Polymers. VSP, Utrecht, 2:137
143. Okamoto K (2003) Proceedings Conf Advances in Materials for Proton Exchange Membrane Fuel Cell Systems, Asilomar, CA, USA, 23–26 February 2003, Prepr 34
144. Rusanov AL, Elshina LB, Bulycheva EG, Müllen K (2003) Polymer Yearbook 18:7
145. Rusanov AL (1992) Russ Chem Rev 61:449
146. Rusanov AL (1994) Adv Polym Sci 111:116
147. Rusanov AL, Elshina LB, Bulycheva EG, Müllen K (1999) Polymer Sci A 41:7
148. USSR Patent 332111 (1972)
149. Kravchenko TV, Bogdanov MM, Kudryavtcev GI, Volokhina AV, Batik'an BA (1972) Khim Volokna 3(14):15
150. Uno K, Niume K, Iwata Ya, Toda F, Iwakura Y (1977) J Polymer Sci 15:1309
151. Goethals ET (1965) Med Vlaame Chem 27:185
152. Goethals ET (1966) Med Vlaame Chem 28:24
153. US Patent 3 536 674 (1970)
154. Dang TD, Venkatasubramanian N, Dean DR, Price GE, Arnold FE (1997) Am Chem Soc Polym Prepr 38(2):301
155. Dang TD, Venkatasubramanian N, Dean DR, Price GE, Arnold FE (1997) Am Chem Soc Polym Prepr 38(2):303
156. Asensio JA, Borros S, Gomez-Romero P (2002) J Polym Sci Polym Chem Ed 40:3703
157. Dong TD, Koerner H, Dalton MJ, Iacobacci AM, Venkatasubramanian N, Arnold FE (2003) Am Chem Soc Polym Prepr 44(1):927

158. Kim S, Cameron DA, Lee Y, Reynolds JR, Savage CR (1999) J Polymer Sci Polym Chem 34:481
159. Sakaguchi Y, Kitamura K, Nakao J, Hamamoto S, Tachimori H, Takase S (2001) PMSE 84:899
160. Einsla BR, Kim YJ, Tchatchowa C, McGrath JE (2003) Am Chem Soc Polym Prepr 44(2):645
161. Gieselman M, Reynolds JR (1990) Macromolecules 23:3188
162. Tarayanagi M (1983) Pure Appl Chem 55:819
163. US Patent 4 814 399 (1989) Chem Abstr 111:1 956 601
164. Dong TD, Arnold FE (1992) Am Chem Soc Polym Prepr 33(1):912
165. Gieselman MB, Reynolds JR (1992) Macromolecules 25:4832
166. Gieselman MB, Reynolds JR (1993) Macromolecules 26:5633
167. Glipa X, Haddad ME, Jones DJ, Roziere J (1997) Solid State Ionics 97:323
168. Kawahara M, Rikukawa M, Sanui K, Ogata N (1998) In: Proceedings of the 6th International Symposium on Polymer Electrolytes. (Extended Abstracts), p 98
169. Kawahara M, Rikukawa M, Sanui K, Ogata N (2000) Solid State Ionics 136:1191
170. Kawahara M, Rikukawa M, Sanui K (2000) Adv Technol 11:544
171. Tsuruhara K, Hara K, Kawahara M, Rikukawa M, Sanui K, Ogata N (2000) Electrochim Acta 45:1223
172. Yoshida H, Hatakeyama T, Hatakeyama H (1990) Polymer 31:693
173. Folk M (1980) Can J Chem 58:1495
174. Lafitte B, Karlsson LE, Jannasch P (2002) Macromol Rapid Commun 23:896
175. Karlsson L, Lafitt B, Jannasch P (2003) In: Proceedings Conf Advances in Materials for Proton Exchange Membrane Fuel Cell Systems, Asilomar, CA, USA, 23–26 February 2003, Poster Abstr 14
176. Pinto MR, Shanze KS (2002) Synthesis 9:1293
177. Tan C, Pinto MR, Shanze KS (2002) Chem Comm 446
178. Pinto MR, Reynolds JR, Shanze KS (2002) Am Chem Soc Polym Prepr 43(1):139
179. Child AD, Reynolds JR (1994) Macromolecules 27:1975
180. Kim S, Jackiw J, Robinson E, Shanze KS, Reynolds JR, Baur J, Rubner MF, Boils D (1998) Macromolecules 31:964
181. Savinell RF, Yeager E, Tryk D, Landau U, Wainright JS, Weng D, Lux K, Litt M, Rogers C (1994) J Electrochem Soc 141:46
182. Wainright JS, Wang J-T, Savinell RF, Litt M, Moadde H, Rogers C (1994) Proc Electrochem Soc 94:255
183. Wainright JS, Wang J-T, Weng D, Savinell RF, Litt M (1995) J Electrochem Soc 142:121
184. Wang J-T, Wasmus S, Savinell RF (1996) J Electrochem Soc 143:233
185. Samms SR, Wasmus S, Savinell RF (1996) J Electrochem Soc 143:225
186. Wainright JS, Savinell RF, Litt MH (1997) In: Proceedings of the Second International Symposium on New Materials for Fuel Cell and Modern Battery Systems, Montreal, Canada, 6–10 July 1997, p 808
187. Kawahara M, Morita J, Rikukawa M, Sanui K, Ogata N (2000) Electrochim Acta 45:1395
188. Jones DJ, Rosiere J (2001) J Membr Sci 185:41
189. Kawahara M, Rikukawa M, Sanui K, Ogata N (2000) Solid State Ionics 136:1193
190. Xiao L, Zhang H, Choe E-W, Scanlan E, Ramanathan LS, Benicewicz BC (2003) In: Proceedings Conf Advances in Materials for Proton Exchange Membrane Fuel Cell Systems, Asilomar, CA, USA, 23–26 February 2003, Prepr 10

191. Powers ED, Serad GA (1986) High Performance Polymers: Origin and Development. Elsevier, New York
192. Stevens JR, Wieczorek W, Raducha D, Jeffrey KR (1997) Solid State Ionics 97:347
193. Wieczorek W, Stevens JR (1997) Polymer 38:2057
194. Tanaka R, Yamamoto H, Kawamura S, Iwase T (1995) Electrochim Acta 40:2421
195. Daniel MF, Destbat B, Cruege F, Trinquet O, Lassegues JC (1988) Solid State Ionics 28:637
196. Agmon N (1995) Chem Phys Letters 244:456
197. Kerres J, Ullrich A, Meier F, Haring T (1999) Solid State Ionics 125:243
198. Kopitzke RW, Lincous CA, Nelson GL (1998) J Polym Sci Polym Chem 36:1197
199. Zaidi SMT, Chen SF, Mikhailenko SD, Kaliaguine S (2000) J New Mater Electrochem Syst 3:27
200. Hou S, Ding M, Gao L (2003) Macromolecules 36:3826

Editor: Manfred Schmidt

Adv Polym Sci (2005) 179: 135–195
DOI 10.1007/b104481
© Springer-Verlag Berlin Heidelberg 2005
Published online: 6 June 2005

Polymer-Clay Nanocomposites

Arimitsu Usuki (✉) · Naoki Hasegawa · Makoto Kato

Toyota Central R&D Labs., Inc., 480-1192 Nagakute, Aichi, Japan
usuki@mosk.tytlabs.co.jp

1	**Introduction**	136
2	**Classifying the Production of Polymer-Clay Nanocomposites**	
	According to the Synthetic Method Employed	137
2.1	Monomer Intercalation Method	137
2.2	Monomer Modification Method	138
2.3	Covulcanization Method	139
2.4	Common Solvent Method	140
2.5	Polymer Intercalation Method	140
3	**Nylon-Clay Nanocomposites**	140
3.1	Clay Organization and Monomer Swelling	140
3.2	Synthesizing the Nylon-Clay Nanocomposite	142
3.3	Characterization of NCH	142
3.4	Properties of NCH	148
3.4.1	Synthesis	148
3.4.2	Mechanical Properties	149
3.4.3	Gas Barrier Characteristics of NCH	150
3.5	Improving the NCH Fabrication Method	151
3.5.1	Characteristics of One-Pot NCH	151
3.5.2	Dry Compound Method	151
3.5.3	Master Batch Method	152
3.5.4	Wet Compound Method	152
3.6	Synthesizing NCH Using Different Types of Clay	153
3.7	Crystal Structure of NCH	155
3.7.1	Alignment of Silicate Layers in NCH	156
3.7.2	Alignment of Nylon 6 Crystals	159
3.8	Other Types of Nylon	162
3.9	Functions of NCH	163
3.9.1	Flame Resistance	163
3.9.2	Self-passivation	163
4	**Polyolefin Clay Nanocomposites**	163
4.1	Introduction	163
4.2	Fabricating Modified Polypropylene-Clay Nanocomposites	165
4.3	Physical Properties of the Modified Polypropylene-Clay Nanocomposites .	169
4.4	Fabricating a Polypropylene-Clay Nanocomposite	
	Using Maleic Anhydride- Modified Polypropylene as a Compatibilizer	
	and Evaluating the Characteristics	173

4.4.1	Effects of the Compatibility between Modified Polypropylene and Polypropylene	173
4.4.2	Effect of the Type of Clay	178
4.5	Fabricating a Polyethylene-Clay Nanocomposite and Evaluating its Physical Properties	180
4.6	Fabricating an Ethylene Propylene Rubber-Clay Nanocomposite and Evaluating the Characteristics	183
4.7	Synthesizing an Ethylene Propylene Diene Rubber (EPDM)-Clay Nanocomposite and Evaluating its Characteristics	189
4.8	Synthesizing a Polyolefin-Clay Nanocomposite Using the Polymerization Method	190
5	**Green Nanocomposites**	192
5.1	Bio-related Polymer-Clay Nanocomposites	192
5.2	Plant Oil-Clay Nanocomposite	192
6	**Conclusion**	193
References		194

Abstract The development of polymer-clay nanocomposite materials, in which nanometer-thick layers of clay are dispersed in polymers, was first achieved about 15 years ago. Since then, the materials have gradually become more widely used in applications such as automotive production. The first practical nylon-clay nanocomposite was synthesized by a monomer intercalation technique; however, the production process has been further developed and a compound technique is currently widely used. A polyolefin nanocomposite has been produced by the compound method and is now in practical use at small volume levels. In this review, which focuses on nylon- and polyolefin-nanocomposites, detailed explanations of production methods and material properties are described. This article contains mainly the authors' work, but aims to provide the reader with a comprehensive review that covers the works of other laboratories too. Lastly, the challenges and directions for future studies are included.

Keywords Clay · Hybrid Intercalation · Nylon · Polyolefin

1
Introduction

An example of a typical material composition is the combination of a polymer and a filler. Because compounding is a technique that can complement the drawbacks of conventional polymers, it has been studied over a long period and its practical applications are well known. Reinforcing materials such as "short-fiber" are often used for compounding with thermoplastic polymers in order to improve their mechanical or thermal properties. Polypropylene and polyamide (nylon) are used for the thermoplastic polymers, while glass fiber and carbon fiber are mainly used as reinforcing materials. A few μm of filler are typically incorporated into the composite materials to improve their properties. The polymer matrix and the fillers are bound together by weak intermolecular forces; chemical bonding is scarcely involved. If the re-inforcing material in the composite can be dispersed on a molecular scale (nanometer level) and can be bound to the matrix by chemical bonding, then significant improvements in the kinetic properties of the material can be achieved or unexpected new properties may be discovered. These are the general goals of polymer nanocomposite studies. In order to achieve this purpose, clay minerals have been discussed as candidates for a filler mate-rial. A layer of silicate clay mineral is about 1nm in thickness with platelets of around 100 nm in width, so it represents a filler with a very large aspect ratio. For comparison, if a glass fiber was 13 μm in diameter with a length of 0.3 mm, it would be 4×10^{-9} times the size of a typical silicate layer. In other words, if the same volumes of glass fiber and silicate were evenly dis-persed, there would be roughly 10^9 times more silicate layers. Furthermore, the specific surface area available would increase exponentially.

A nylon-clay hybrid (nanocomposite NCH) was originally developed by the authors and was the first polymer nanocomposite to be used practically. Since 1990, when it was first used, various studies and analyses of it have been reported. An excellent review was published in 2003 [1]. In the present review, which focuses on the authors' studies, details on the NCH that we re-ported initially and further developments in polypropylene and polyethylene will be described. In Sect. 2, comprehensive classifications of the production methods developed previously will be described, according to the synthesis method employed. Thereafter, nylon will be discussed in Sect. 3, polyolefin in Sect. 4, and renewable polymer (green polymer) will be discussed in Sect. 5.

2
Classifying the Production of Polymer-Clay Nanocomposites According to the Synthetic Method Employed

2.1
Monomer Intercalation Method

In this method, clay is first ion-exchanged using an organic compound in order for the monomer to be intercalated into the layers of the clay. The monomers that form the intercalated layer will become a polymerized inter-layer. The authors succeeded in producing a polyamide nanocomposite for the first time using this method. The details will be described in Sect. 3. The basic concept of the technique is as follows.

A polymerization to produce nylon 6 proceeds via the ring-opening poly-merization of ε-caprolactam. This can occur in the presence of clay after

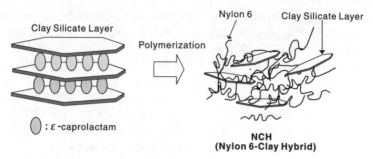

Fig. 1 Schematic diagram of polymerization to NCH

ε-caprolactam is intercalated into a clay gallery such that the silicate layers are dispersed uniformly in the nylon 6 matrix. It was found that organophilic clay that had been ion-exchanged with 12-aminododecanoic acid could be swollen by molten ε-caprolactam (the basal spacing expanded from 1.7 nm to 3.5 nm) [2]. ε-Caprolactam was polymerized in the clay gallery and the silicate layers were dispersed in nylon 6 to yield a nylon 6-clay hybrid (NCH) [3]. This is the first example of an industrial clay-based polymer nanocomposite. Figure 1 shows a schematic representation of the polymerization.

The modulus of NCH increased to 1.5 times that of nylon 6, the heat distortion temperature increased to 140 °C from 65 °C, and the gas barrier effect was doubled at a low loading (2 wt %) of clay [4].

There is another example, in which ε-caprolacton is polymerized in a clay gallery in the same manner. In this case, the gas permeability decreased to about 20% under 4.8 vol % (12 wt %) of clay addition [5]. There is yet another example of an epoxy resin-clay nanocomposite. In this case, the tensile strength and modulus increased drastically upon the addition of 2–20 wt % clay [6].

2.2
Monomer Modification Method

This is a method in which an acryl monomer is modified in a layer prior to the polymerization of the acrylic resin.

In one example, a quaternary ammonium salt of dimethylaminoacrylamide (Q: modified monomer of acrylic resin) was ion-bonded to silicate layers, while ethyl acrylate (EA) and acrylic acid (Aa) were copolymerized in the clay gallery. The ratio between the EA and the Aa was 10 : 1 (molar ratio). Four kinds of acrylic resin-clay nanocomposites were polymerized. Their clay contents were 1, 3, 5 and 8 wt % on the basis of the solid acrylic resins. Suspensions with greater than 3 wt % clay addition acted as pseudoplastic fluids. Transparent acrylic resin-clay nanocomposite films cross-linked by melamine

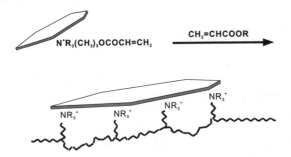

Fig. 2 Scheme of acrylic resin-clay nanocomposite

were formed, and the gas permeability of the films decreased to about 50% under 3 wt % clay addition [7].

There are also some other reports of acrylic resin-clay nanocomposites. A poly(methylmethacrylate) clay nanocomposite was synthesized using a modified organophilic clay in the same manner [8], and by emulsion polymerization [9]. Figure 2 shows a schematic representation of this polymerization method.

2.3
Covulcanization Method

The basal spacing of an organophilic clay ion-bonding nitrile rubber (NBR) oligomer incorporating telechelic amino groups was expanded by 0.5 nm from its initial spacing (1.0 nm) [10]. After this, high molecular weight NBR was kneaded with this organophilic clay and vulcanized with sulfur. It produced an NBR-clay nanocomposite consisting of dispersed clay and co-vulcanized high molecular weight NBR and NBR oligomer [11]. Its permeability to hydrogen and water decreased to 70% on adding 3.9 vol % clay [12].

Figure 3 shows a schematic representation of this production method.

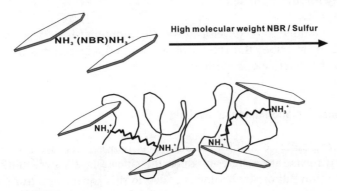

Fig. 3 Scheme of NBR-clay nanocomposite

2.4
Common Solvent Method

In the case of the synthesis of polyimide, the polymerization solvent used for polyamic acid (a precursor of polyimide) is usually dimethyl acetoamide (DMAC). We found that clay ion-exchanged dodecyl ammonium ions could be homogeneously dispersed in DMAC. A solution of this organophilic clay and DMAC was added to a DMAC solution of polyamic acid. The film was cast from a homogeneous mixture of clay and polyamic acid, and was heated at 300 °C to achieve the desired polyimide clay nanocomposite film. Its permeability to water decreased to 50% upon addition of 2.0 wt % clay [13]. It was confirmed that its permeability to carbon dioxide also decreased by half [14].

2.5
Polymer Intercalation Method

Polypropylene (PP)-clay hybrids cannot be easily synthesized because PP is hydrophobic and has poor miscibility with clay silicates. Octadecyl ammonium ions were used as modifiers for the clay, and a polyolefin oligomer was used so that the clay became more compatible. Organophilic clay, a polyolefin oligomer and PP were blended using an extruder at 200 °C. It was confirmed by transmission electron microscopy (TEM) that the clay was dispersed in a monolayer state in the PP matrix. Thus, PP was directly intercalated into the clay gallery [15].

There is also a direct intercalation process in which PP is modified using maleic anhydride, followed by melt compounding [16]. This is a useful process from an industrial standpoint.

3
Nylon-Clay Nanocomposites

The first technique to be developed was the monomer intercalation method. This chapter describes the results of studies and experiments that we conducted to characterize this method.

3.1
Clay Organization and Monomer Swelling

If montmorillonite containing sodium ions between its layers is dispersed in water, it turns into a state in which the silicate layers swell uniformly. If the ammonium salt of alkylamine is added to this aqueous mixture, then the alkylammonium ions are exchanged with the sodium ions. As a result of this reaction, a clay forms in which the alkylammonium ions are intercalated between the layers. Because the silicate layers in the clay are negatively charged, they bond with the alkylammonium ions through ionic bonding if an ammonium salt is injected. If the length and type of the alkyl chain are changed, the hydrophilic and hydrophobic characteristics and other characteristics of this organophilic clay can be adjusted such that surface modification of the clay becomes possible.

A novel compounding technique was developed to synthesize nylon 6 in a clay gallery by modifying the clay surface and intercalating monomers between the clay gallery. The organic materials that are required to synthesize nylon 6 in a clay gallery through the surface modification of the clay must satisfy the following three requirements:

1. The organic material must have an ammonium ion at one end of the chain so that it can bond with clay through ionic bonding.
2. It must have a carboxyl group ($-$ COOH) at the other end to react with ε-caprolactam, which is a nylon 6 monomer, for ring opening and polymerization.
3. It must be polarized in such a manner as to allow the silicate layers to swell in the ε-caprolactam.

It was found that 12-aminododecanoic acid ($H_2N(CH_2)_{11}COOH$) meets all of these requirements [2].

Using a homomixer, 300 g of montmorillonite was uniformly dispersed in 9 l deionized water at 80 °C. 154 g of 12-aminododecanoic acid and 72 g of concentrated hydrochloric acid were added to 2 l deionized water, and they were dissolved at 80 °C. This hydrochloric acid/salt water solution of 12-aminododecanoic acid was mixed with the deionized water in which the montmorillonite was dispersed and the mixture was stirred for five minutes. The mixture was filtered to obtain aggregates, and the obtained aggregates were washed twice with water at 80 °C. They were then freeze-dried. In this way, organophilic clay was obtained in the form of a fine white powder. This organophilic clay was called "12-Mt."

12-Mt and ε-caprolactam were well mixed in a mortar in a weight ratio of 1 : 4, and they were then dried and dehydrated for 12 hours in a vacuum desiccator containing phosphorous pentoxide. These specimens were left in a temperature-controlled bath kept at 100 °C for one hour to cause the ε-caprolactam substances to swell. They were subjected to X-ray diffraction measurements at 25 °C and 100 °C. It was found that two distinct sizes were present at the different temperatures: 3.15 nm (25 °C) and 3.87 nm (100 °C) and that the one processed at 100 °C had caprolactam molecules intercalated between the layers.

3.2
Synthesizing the Nylon-Clay Nanocomposite

The mixture was placed in a glass reaction container of about 500 ml volume and then dehydrated and deoxidized under a reduced pressure (5×10^{-2} torr). The glass reaction container was then sealed. The mixture in this container was further heated at 120 °C for 12 hours and at 250 °C for 48 hours to polymerize the ε-caprolactam. 12-Mt was added in weight percentages of 2, 5, 8, 15, 30, 50 and 70. After polymerization was complete, the mixture was taken out of the reaction container and pulverized using a Fitz mill. The pulverized materials were washed with water at 80 °C, and any residual monomers and low-molecular weight compounds were removed. They were further dried for 12 hours at 80 °C in a vacuum to obtain NCH. The loadings of 12-Mt were expressed by wt %, and NCHs for each different loading of 12-Mt were called NCH2, NCH5, and so on to NCH70.

Additional specimens were prepared by melting and kneading sodium-type montmorillonite (unorganized type) and nylon 6 using a twin screw extruder at 250 °C for the purpose of comparing them with the specimens prepared as described above. This method of preparing specimens is commonly used when compounding particulate fillers with polymers. The composite material prepared in this way was called NCC (Nylon 6-Clay Composite), and the NCC was compared with the NCH.

3.3
Characterization of NCH

Figure 4 shows the X-ray diffraction spectra. With NCH70 and NCH50, a clear peak showing the interlayer distance associated with the $d(001)$ plane of montmorillonite was observed. With NCH30 and NCH15, however, the peak

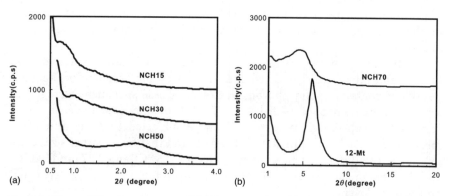

Fig. 4 (a) X-ray diffraction patterns of NCH15, 30 and 50 (b) X-ray diffraction patterns of NCH70 and 12-Mt

Table 1 Basal spacing for NCHs

	Content of Clay (wt %)	Basal spacing from XRD
NCH2	1.5	...
NCH5	3.9	...
NCH8	6.8	...
NCH15	13.0	12.1
NCH30	26.2	6.0
NCH50	42.8	4.4
NCH70	59.6	2.6
12-Mt	78.7	1.7
Nylon 6	0	–

was weak and took the form of a shoulder. With NCH2, NCH5 and NCH8, no peak was observed in the measurement range. The point where the inflection rate reaches a maximum in the shoulder-shaped spectrum was defined as the peak of $d(001)$ to calculate the interlayer distance. The results are shown in Table 1 [3].

Figure 5 shows the surfaces of press-molded NCH and NCC products. The surface of the press-molded NCH product is smooth, whereas many aggregates (clay minerals) on a millimeter-scale were observed on the surface of the press-molded NCC product. Furthermore, many bubbles were observed during the molding of the press-molded NCC product. This is thought to be due to the effects of water contained in the sodium-type montmorillonite.

To observe the dispersed state of silicate layers in the NCH more closely, the press-molded NCH product was observed using a TEM at high magnification. The results of this observation are shown in Fig. 6. As shown in this figure, the cross-sections of the silicate layers have a black, fibrous appear-

NCC(clay content : 5wt%)   NCH(clay content : 4.2wt%)

Fig. 5 Surface appearances of NCC and NCH

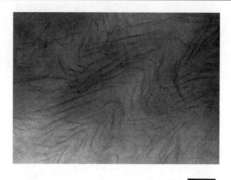

100nm

Fig. 6 Transmission electron micrograph of a section of NCH

ance, and the silicate layers are uniformly dispersed at a molecular level in nylon 6. It was found that the interlayer distances of the NCH15 and NCH30, as measured by X-ray diffraction, are nearly equal to those measured by TEM.

The relationship between the interlayer distance ds in the silicate layers and the amount of 12-Mt in the NCH was analyzed as follows. Providing that the ratio of the amount of nylon 6 to the amount of 12-Mt is R, then Eq. 1 holds true:

$$R = \varrho n \cdot (ds - t)/\varrho c \cdot t \tag{1}$$

in which R is nylon 6/12-Mt (g/g), ϱn is the concentration of nylon 6 (1.14 g/cm^3), ϱc is the concentration of 12-Mt (1.9 g/cm^2), and t is the interlayer distance of 12-Mt (1.72 nm).

If each of the symbols in Eq. 1 is replaced by the numerical values, we have:

$$ds = 2.87R + 1.72 . \tag{2}$$

Figure 7 shows the ds values calculated using Eq. 2, as well as actual measurements. The actual measurements are slightly lower than the calculated values. These results show that each silicate layer is dispersed in nylon 6. The fact that the actual measurements differ from the calculated values indicates that nylon exists not only inside, but also outside the layers. The ratio pi of the nylon inside the layers to the nylon inside and outside the layers can be calculated using Eq. 3:

$$pi = (d_0 - 1/d_c - 1) \times 100 . \tag{3}$$

in which d_0 is the observed interlayer distance and ds is the interlayer distance calculated using Eq. 2.

The pi value of NCH15 was 73.0% and that of NCH70 was 97.6%. The π value increased as the amount of 12-Mt increased.

These results show that 12-Mt initiates the ε-caprolactam polymerization and that most of the nylon is polymerized between the 12-Mt layers.

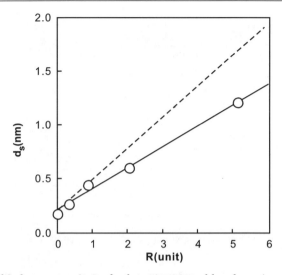

Fig. 7 Relationship between ratio R of nylon 6/12-Mt and basal spacings. (d_s) *Solid line*: observed value, *dotted line*: calculated value from Eq. 2

Table 2 shows the results of measurements of the amount of montmorillonite and the amount of terminal groups in each NCH. Figure 8 shows a graph in which the concentration of terminal groups is plotted relative to the amount of montmorillonite. If the amount of montmorillonite increased, the concentration of carboxyl groups increased almost linearly, while the amount of amino groups remained almost unchanged. In addition, the amount of carboxyl groups far exceeded the amount of amino groups in each NCH. This is thought to be attributed to the fact that the ions of some amino

Table 2 End group analysis results for NCHs

	Content of Clay (wt %)	C_{NH_2}	C_{COOH}	Mn from C_{COOH}
		from end group analysis (10^{-5} eq/g)		(10^3)
NCH2	1.5	3.85	5.69	17.2
NCH5	3.9	4.86	9.49	10.0
NCH8	6.8	6.70	14.4	6.34
NCH15	13.0	8.04	22.9	3.80
NCH30	26.2	12.6	44.3	1.66
NCH50	42.8	12.1	70.6	0.810
NCH70	59.6	6.64	86.7	0.466
12-Mt	78.7	–	–	(0.216)[a]
Nylon 6	0	5.69	5.41	–

[a] Molecular weight of 12-aminolauric acid

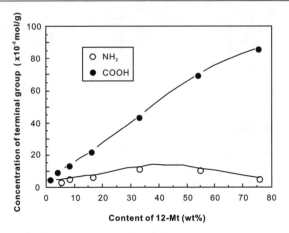

Fig. 8 Relationship between 12-Mt content and end group concentration

groups at the N-end of the nylon molecules combine with the ions of the silicate layers of the montmorillonite to form ammonium ions.

If the montmorillonite content is Wm (wt %), the amount $C_{NH_3}+$ (mol/g) of ammonium groups in NCH can be calculated based on the equivalence between the montmorillonite's cation exchange capacity (CEC) and the ammonium groups, as shown in Eq. 4 below:

$$C_{NH_3}+ = Wm \times CEC/100 \tag{4}$$

In this equation, CEC is 1.2×10^{-3} eq/g. The relationships between the amino, carboxyl and ammonium groups are defined based on the equivalence between the N- and C-ends of the nylon 6 molecules, as shown in Eq. 5 below:

$$C_{NH_3}+ + C_{NH_2} = C_{COOH} \tag{5}$$

in which C_{NH_2} is the amount (mol/g) of amino groups and C_{COOH} is the amount (mol/g) of carboxyl groups. Therefore, Eq. 6 can be formulated from Eqs. 4 and 5 as follows:

$$C_{NH_3} = C_{COOH} - C_{NH_2} = Wm \times 1.2 \times 10^{-5} . \tag{6}$$

Table 3 shows the values calculated using Eq. 6 and the measured values ($C_{COOH} - C_{NH_2}$). As is apparent from Table 3, both values are in good agreement. This shows that the N-end of the nylon 6 turns into an ammonium group and the ions in the ammonium group combine with the ions in the montmorillonite layers. The number average molecular weight (Mn) of the nylon 6 is expressed as the inverse number of the mole number per gram of nylon 6. The number average molecular weight Mn of nylon 6 in NCH can be calculated based on the amount C_{COOH} of end carboxyl groups and the

Table 3 Calculated anion site number for clay and observed values of $C_{COOH} - C_{NH_2}$

	$C_{NH_3^+}$	$C_{COOH} - C_{NH_2}$ $(10^{-5}$ eq/g)
NCH2	1.79	1.84
NCH5	4.64	4.60
NCH8	8.09	7.69
NCH15	15.5	14.9
NCH30	31.2	31.7
NCH50	50.9	58.5
NCH70	70.9	80.1

montmorillonite content Wm, as shown in Eq. 7 below:

$$Mn = 1/\{C_{COOH}[100/(100 - Wm)]\}. \tag{7}$$

Table 2 shows the results of calculations made using this equation.

The molecular weight decreased as the amount of 12-Mt increased. Assuming that the carboxyl group in 12-Mt is the only active site and that the polymerization reaction progresses without side reactions, the molecular weight Mn can be expressed by Eq. 8:

$$Mn - 216 = (1/Cm) \times (1 - f)/f \times p \tag{8}$$

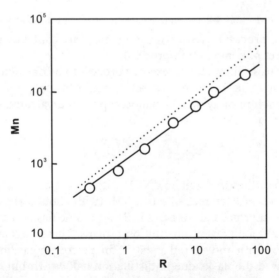

Fig. 9 Relationship between ratio R of nylon 6/12-Mt and molecular weight (Mn). *Solid line*: observed value, *dotted line*: calculated value from Eq. 9

in which Cm is the amount of carboxyl groups in 12-Mt (9.6×10^{-4} mol/g), f is the wt % of injected 12-Mt, p is the caprolactam inversion rate (%), and 216 is the molecular weight of 12-aminododecanoic acid.

Because $(1 - f) \cdot p/f = R$, Eq. 8 can also be expressed as follows:

$$Mn = 1.04 \times 103 \times R + 216 \tag{9}$$

Mn values calculated using Eq. 9 as well as the measured Mn values are shown in Fig. 9.

The gradient of the measured Mn values is smaller than that of the calculated values. This means that there were active sites other than that of the end carboxyl group in the 12-Mt during polymerization.

The amount of other active sites was estimated from the gradient of the graph in Fig. 9, and was found to be 4.8×10^{-4} mol/g. This means that about 0.8 wt % of water is contained in 12-Mt, and this weight percent figure was supported by measured values.

3.4
Properties of NCH

The compound with up to 8 wt % of 12-Mt (NCH8) could be molded using a 40 ton injection molder to make a test specimen. NCHs were prepared by adding 2, 5 and 8 wt % of 12-Mt, and these NCHs were polymerized. NCH with more than 8 wt % of 12-Mt had very poor flow properties, and could not be molded. Therefore, the mechanical properties of this NCH could not be measured.

Although 12-Mt was freeze-dried and used for analytical processing (dry polymerization process), 12-Mt with some moisture content was used in this experiment (wet polymerization process).

This wet process allows the freeze-dry process to be omitted and the polymerization time to be shortened. Therefore, it has the potential to be used for the mass-production of nylon clay nanocomposites on a commercial basis.

3.4.1
Synthesis

This section describes in detail how NCH with 5 wt % of 12-Mt was polymerized. 509 g of ε-caprolactam, 29.7 g of 12-Mt (with about 300 g of water), and 66 g of 6-aminocaproic acid were put in 31 separable flasks with stirrers, and were subjected to a nitrogen substitution process. These flasks were then immersed in an oil bath and stirred at 250 °C in a nitrogen gas flow for 6 hours. Water overflowed the flasks due to distillation halfway through this process. For the NCHs with 2, 5 and 8 wt %, polymerization was terminated when the load on the stirrers increased to a certain level.

After the flasks were cooled, aggregated polymers were removed from the flasks and pulverized. They were then washed with water at 80 °C three times, and any monomers and oligomers that remained unreacted were removed. A series of NCH specimens were obtained in this way.

These specimens were labeled NCH2, NCH5 and NCH8, according to their loadings of 12-Mt. Nylon 6 "1013B" (molecular weight: 13 000) made by Ube Industries, Ltd. was used as a specimen that contained no montmorillonite.

3.4.2
Mechanical Properties

Table 4 shows the mechanical properties of NCH against those of nylon 6 (1013B). As is apparent from Table 4, NCH is superior to nylon 6 in terms of its strength and elasticity modulus. In the case of NCH5 in particular, the tensile strength at 23 °C is 1.5 times higher than that of nylon 6, the bending strength at 120 °C is twice that of nylon 6, and the flexural modulus at 120 °C is about four times as large as that of nylon 6. However, its impact strength is below that of nylon 6.

The heat distortion temperature of NCH5 increased to 152 °C, and the heat resistance also improved. Figure 10 shows the heat distortion temperatures relative to the clay content. As shown in Fig. 10, the values indicate that the clay is in an almost saturated state in NCH5 [4].

Table 4 Properties of NCH and Nylon 6

Properties		Unit	NCH2	NCH5	NCH8	Nylon 6
Tensile	23 °C	MPa	76.4	97.2	93.6	68.6
strength	120 °C		29.7	32.3	31.4	26.6
Elongation	23 °C	%	> 100	7.30	2.5	> 100
	120 °C		> 100	> 100	51.6	> 100
Tensile	23 °C	GPa	1.43	1.87	2.11	1.11
modulus	120 °C		0.32	0.61	0.72	0.19
Flexural	23 °C	MPa	107	143	122	89.3
strength	120 °C		23.8	32.7	37.4	12.5
Flexural	23 °C	GPa	2.99	4.34	5.32	1.94
modulus	120 °C		0.75	1.16	1.87	0.29
Charpy impact strength (without notch)		kJ/m^2	102	52.5	16.8	> 150
Heat distortion temperature		°C	118	152	153	65

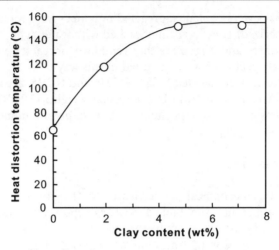

Fig. 10 Dependence of heat distortion temperature on clay content

The characteristics of the dependency of injection-molded NCH products on layer thickness have been investigated and reported [17]. An investigation of 0.5, 0.75, 1.0 and 2.0 mm-thick test specimens revealed that the thicker the product, the lower the elastic modulus under tension becomes.

3.4.3
Gas Barrier Characteristics of NCH

Table 5 shows a comparison between the gas barrier characteristics of NCH (with 0.74 vol % of montmorillonite) and those of nylon 6. The hydrogen permeability and water vapor permeability coefficients of NCH with only 0.74 vol % of montmorillonite were less than 70% of the equivalent coefficients for nylon 6, indicating that NCH has superior gas barrier characteristics.

This gas barrier effect of NCH can be explained by postulating that the added fillers caused the diffusion paths of the gases to meander, such that the gases were forced to follow complicated, meandering paths, and hence the diffusion efficiency decreased.

When gas travels through NCH, the permeability coefficient of the gas can be analyzed using a geometrical model in which silicate layers are dispersed. In NCH, silicate layers are aligned nearly parallel with the film surface. According to Nielsen, the diffusion coefficient D of a liquid or a gas can be calculated using Eq. 10 if the liquid or gas is in a composite material in which plate particles are in a planar orientation:

$$D = D_0/\{1 + (L/2d)V\} \qquad (10)$$

Table 5 Permeability of NCH and Nylon 6

Permeability	NCH*	Nylon 6
Permeability of hydrogen $\times 10^{-11}$ /cm^3 (STP) cm cm^{-2} s^{-1} cm Hg^{-1}	1.79	2.57
Permeability of water vapor $\times 10^{-10}$ /g cm cm^{-2} s^{-1} cm Hg^{-1}	1.78	2.83

* 0.74 vol %

where D_0 is the diffusion coefficient in a matrix, L is the size of one side of a plate particle, d is the thickness, and V is the volume fraction of particles.

Providing that L is 100 nm, d is 1 nm, and V is 0.0074, we have $D/D_0 = 0.73$.

This value is equivalent to both 0.70, which is the experimental value obtained for hydrogen, and 0.63, which is the experimental value obtained for water. This shows that the gas barrier characteristics of NCH should be interpreted as being due to the geometrical detour effect of the silica layers of the montmorillonite.

3.5
Improving the NCH Fabrication Method

3.5.1
Characteristics of One-Pot NCH

In addition to the method of organizing clay and then adding monomers for polymerization, the "one-pot" polymerization method has also been proposed [18]. By mixing montmorillonite, caprolactam and phosphoric acid simultaneously in a container and polymerizing them, NCH can be produced quite readily. The dispersed state of the clay mineral and the mechanical properties of specimens produced in this way are the same as those of specimens made by the polymerization method. NCH was successfully synthesized by the one pot technique, and so the process time could be shortened.

3.5.2
Dry Compound Method

Besides the polymerization method, a method of directly mixing nylon polymers and organophilic clay using a twin screw extruder was developed. Although clay mineral is not dispersed sufficiently using a single screw extruder (screw speed: 40 rpm, barrel temperature: 240 °C), it can be well-dispersed using a twin screw extruder (screw speed: 180 rpm, barrel temperature: 240 °C). Experimental results and mechanical characteristics have been reported by Toyota CRDL, Allied Signal and the Chinese Academy of Sciences with reference to benchmarking values [19, 20].

3.5.3
Master Batch Method

To produce composite materials on a commercial basis, a master batch method of diluting materials and mixing them in specified proportions is widely used. A case is known in which this method was used to prepare nylon clay nanocomposites. If high molecular weight grades (Mn= 29 300) of nylon 6 are used, the level of exfoliation of clay becomes higher than when low molecular weight grades (Mn= 16 400) of nylon 6 are used. To minimize the clay exfoliation, nylon 6 of high molecular weight was mixed with 20.0, 14.0 and 8.25% of clay to prepare the master batches. Each master batch of nylon 6 mixed with clay was diluted using nylon 6 of low molecular weight. The mechanical properties of the nylon 6-clay nanocomposite prepared in this way were found to be almost the same as the mechanical properties of nylon 6 of high molecular weight produced using the dry compound method by the addition of 6.5, 4.0 and 2.0% of clay [21, 22].

3.5.4
Wet Compound Method

The process of organizing clay using ammonium ions has a considerable impact on the production cost. In order to omit this process, silicate layers of clay (sodium-type montmorillonite) that are uniformly dispersed in water were turned into slurry form and mixed with a molten resin. The concept of this method is shown in Fig. 11. Clay slurry was injected using first a twin screw extruder and then a screw feeder, and water was removed under reduced pressure. In this process, a nanocomposite consisting of nylon and clay minerals uniformly dispersed in nylon was successfully fabricated. This method makes the simplification of the clay organization process possible,

Table 6 Properties of NCH

Specimen	Clay content (%)	Tensile Strength (MPa)	Tensile Modulus (GPa)	Heat Distortion Temperature (°C at 18.5 kg/cm)
Nylon 6	0	69	1.1	75
Synthesized NCH	1.9	76	1.43	118
Dry compounding NCH	1.8	82	1.41	135
Clay slurry compounding NCH	1.6	82	1.38	102

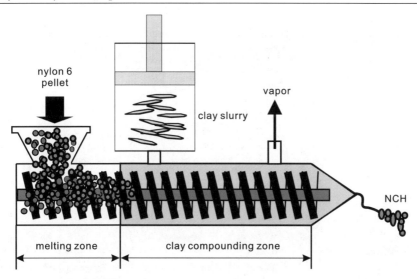

Fig. 11 Schematic figure depicting the compounding process for preparing nanocomposites using the clay slurry

with the advantage that nanocomposites can be obtained at low cost. Table 6 shows the mechanical properties of this nanocomposite. The heat distortion temperature dropped somewhat because the bonding of clay and nylon does not occur by ionic bonding [23].

3.6
Synthesizing NCH Using Different Types of Clay

Other than montmorillonite, synthetic mica, saponite and hectorite were used to synthesize a nylon 6-clay hybrid. The nanocomposites fabricated by using each of these types of clay were called NCH, NCHM, NCHP and NCHH.

Silicate layers were uniformly dispersed in nylon 6 in NCH, NCHM, NCHP and NCHH at the molecular level. The thicknesses of the silicate layers were 1 nm in all of these nanocomposites, but their widths varied depending on the type of clay used. An examination of each photograph revealed that the width of the nanocomposites fabricated using montmorillonite and synthetic mica were about 100 nm and those of the nanocomposites fabricated using saponite and hectorite were about 50 nm.

Table 7 shows the mechanical properties of each nanocomposite. The tensile strengths of each nanocomposite at 23 °C and 120 °C are as follows:

NCH (montmorillonite) > NCHM (synthetic mica)

$$> NCHP(saponite) \geq NCHH (hectorite) .$$

Table 7 Properties of NCH synthesized using 5 wt % organic clay

Properties Clay		NCH montmo-rillonite	NCHM mica	NCHP saponite	NCHH hectorite	Nylon 6 none
Tensile strength (MPa)	23 °C	97.2	93.1	84.7	89.5	68.6
	120 °C	32.3	30.2	29.0	26.4	26.6
Elongation	23 °C	7.3	7.2	> 100	> 100	> 100
Tensile modulus (GPa)	23 °C	1.87	2.02	1.59	1.65	1.11
	120 °C	0.61	0.52	0.29	0.29	0.19
Heat distortion temperature (°C)		152	145	107	93	65
Heat of fusion (J/g)		61.1	57.2	51.5	48.4	70.9
Heat of fusion (J/nylon 6 1 g)		63.6	59.6	53.4	50.4	70.9

The heat distortion temperatures of each nanocomposite are as follows:

NCH > NCHM > HCHP > NCHH

To check the differences between the mechanical properties of these nano-composites, the interface affinity between clay and nylon 6 was analyzed by measuring the NMR of nitrogen at the chain end in nylon 6. Because the concentration of nitrogen at the chain end in nylon 6 is extremely low, glycine (H_2NCH_2COOH) and hexamethylene diamine ($H_2N(CH_2)_6NH_2$) were used as model compounds.

Table 8 shows the 15N chemical shift of glycine-organized clay and hexamethylene diamine (HMDA). Because glycine contains ampholite ions in the neutral state, the HMDA values were used as the chemical shift values of neutral N.

The ^{15}N chemical shifts of four types of glycine-organized clays were found to occur midway between the most polarized glycine hydrochloride (15.6 ppm) and neutral HMDA (7.0 ppm).

As the chemical shift moved toward lower fields, its electron density decreased. This means that nitrogen was polarized more toward the positive direction ($\delta +$). It is thought that if $\delta +$ of nitrogen is large, stronger ionic bonding with the negative charge of the silicate layers of the clay can be realized. Montmorillonite in four types of clay had the largest $\delta +$, 11.2 ppm. $\delta +$ decreased in the order of synthetic mica > saponite $\geq$ hectorite.

It was inferred from all of these results that montmorillonite in all types of clay can bond most strongly with nylon 6 and that the bond strength weakens in the order of synthetic mica > saponite $\geq$ hectorite. Figure 12 shows

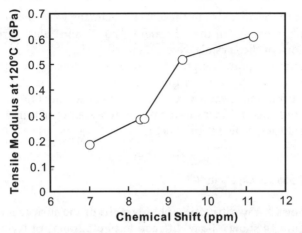

Fig. 12 Relation between N-NMR chemical shifts of model compounds and tensile modulus of nylon 6 clay nanocomposites at 120 °C

Table 8 ^{15}N-NMR chemical shifts of model compounds

Compounds	Chemical shift* (ppm)		
$Cl^-NH_3^+CH_2COOH$	15.6	ionized	large
Montmorillonite-$NH_3^+CH_2COOH$	11.2		
Mica-$NH_3^+CH_2COOH$	9.4	partialy	δ^+ on
Saponite-$NH_3^+CH_2COOH$	8.4	ionized	nitrogen atom
Hectorite-$NH_3^+CH_2COOH$	8.3		
HMDA	7.0	neutral	small

* ppm relative to $^{15}NH_4NO_3$

the ^{15}N-NMR chemical shift as an indicator of bond strength and the flexural modulus at 120 °C as the central characteristic value. As is apparent from this figure, close correlations between the chemical shift and the flexural modulus are noted [24].

3.7
Crystal Structure of NCH

The surfaces of NCH and the nylon 6 test specimens were scraped around the center to a depth of 0.5 mm. The surfaces of other NCH and nylon 6 test specimens (3 mm thick) were scraped to a depth of 1 mm. X-ray diffraction photographs of these test specimens were taken using Laue cameras. Specifically, the surfaces and insides of these test specimens were subjected to X-ray diffraction photography in the "through", "edge" and "end" directions, and the orientations of the crystals were examined [25].

The X-ray diffraction strength of these test specimens was also measured using a reflection method. By scraping their surfaces to specified thicknesses, their X-ray diffraction spectra were measured at each thickness. This process of scraping and spectral measurement was repeated to obtain X-ray diffraction spectra at each different thickness.

3.7.1
Alignment of Silicate Layers in NCH

Figure 13 shows X-ray diffraction photographs of the surface and the inside of NCH. Figure 14 shows X-ray diffraction photographs of Nylon 6. "Thru" is a diffraction photograph taken by introducing the X-rays perpendicular to the molded surface. "Edge" is a diffraction photograph taken by introducing the X-rays parallel with the molded surface and perpendicular to the direction of flow on the molded surface. "End" is a diffraction photograph taken by introducing the X-rays in the direction of flow on the molded surface.

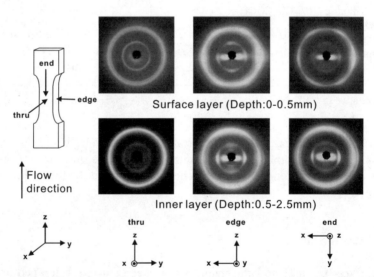

Fig. 13 X-ray diffraction photographs for the surface and inner of an injection-molded NCH bar 3 mm thick. Surface and inner layers correspond to the regions of 0–0.5 mm and 0.5–2.5 mm from the bar surface, respectively. The diffraction photographs are termed thru-, edge-, and end-view patterns, when the X-ray beam was incident on the NCH bar along the x-, y-, and z-axes, respectively, which are also defined in the figure

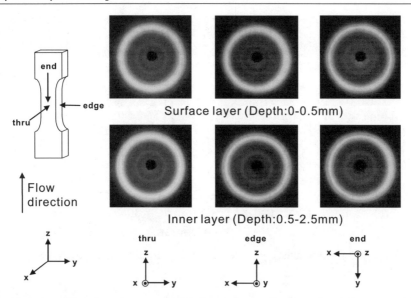

Fig. 14 X-ray diffraction photographs for the surface and inner of an injection-molded Nylon 6 bar 3 mm thick. For photographical conditions, see the legend of Fig. 13

In these figures, x and y represent the directions perpendicular and parallel to the surface of the test specimen. y and z represent the directions perpendicular and parallel to the flow of resin.

In the "end" and "edge" patterns on the surface of the NCH and inside the NCH, a pair of clear streak diffractions are observed in the horizontal direction (x-direction). This shows that the silicate layers are aligned parallel to the molded surface. On the surface of the NCH and inside the NCH, the inside streak of the "end" pattern becomes a little wider toward the azimuth angle. This shows that the alignment of the silicate layers is less orderly inside the NCH than on the surface.

X-ray scattering measurements were made along the x-direction of the "edge" pattern in the surface layers. The diffraction spectrum obtained from this measurement is shown in Fig. 15. The strong scattering peak ($2\theta = 25°$) is thought to be associated with the superposition of the γ-type planes (020 and 110) of nylon 6. On the other hand, the curve that appears between $2\theta = 4°$ and $10°$ is thought to be associated with the clearly-visible streak running from the silicate layers of the montmorillonite. The strength of the clearly-visible streak in the center in Fig. 13 is $2\theta = 10°$, which is at almost the same level as the background. The angle 2θ can be explained based on the hypothesis that 1 nm silicate layers are aligned parallel with the surface of the molded specimen.

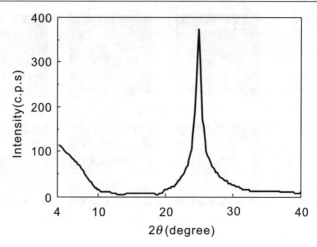

Fig. 15 X-ray diffraction intensity curve along the x-direction for the edge-view patterns of the surface layers in Fig. 13

The strength function $I(q)$ of thin layers (thickness: d) is proportional to the cross-sectional shape:

$$I(q) = Nn_e^2[\sin(qd/2)/(qd/2)] \tag{11}$$

in which q is $4\pi \sin\theta/\lambda$, λ is the wavelength of the X-ray, N is the number of silicate layers aligned in parallel with the surface of the test specimen in the volume irradiated by the X-ray, and n_e is the number of electrons in the silicate layers.

In Eq. 11, the scattering intensity is 0, which is calculated by $q = 2\pi/d$. Providing that this scattering intensity corresponds to the critical value, 2θ, we have the following:

$$\theta = \arcsin(\lambda/2d). \tag{12}$$

By substituting $\lambda = 0.1790$ nm and $d = 1$ nm into Eq. 12, we obtain $2\theta = 10.3°$, which is approximately consistent with the results of this experiment. This shows that silicate layers of 1 nm in thickness (single layers) are dispersed.

It is thought from Eq. 11 that the streak intensity is proportional to the amount of silicate layers that exist in parallel with the surface of a test specimen. Figure 16 shows the relationship between the intensity ($I(4°)$) at $2\theta = 4°$ and the depth from the surface of the NCH test specimen.

The intensity $I4°$ decreases linearly as the depth increases. It becomes almost constant at a point between 0.8 mm and 1.2 mm. After this point, it starts decreasing again. This means that the amount of silicate layers parallel to the surface of a molded specimen continuously decrease in the depth direction. That is, the fluctuations in the silicate layers aligned in the same di-

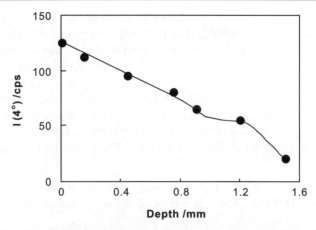

Fig. 16 Scattering intensity, $I(4°)$, of the streak due to the silicate monolayers parallel to the bar surface at a scattering angle of $4°$ as a function of depth from the bar surface. The streak is in the x-direction of the edge-view pattern in Fig. 13

rection as the flow of the resin increase as the depth increases. It is estimated from the inside "end" pattern shown in Fig. 13 that the maximum intensity of this fluctuation is $\pm15°$.

The decreased scattering intensity around the center of a molded specimen is thought to be due to disturbances in the alignment caused by silicate layers that are uniaxially-aligned along the flow direction.

3.7.2
Alignment of Nylon 6 Crystals

The other reflection patterns (except for the reflections off the silicate layers) shown in the diffraction photographs in Fig. 13 are directly related to the γ-type crystals of nylon 6. There have been some previous reports concerning the γ-type crystal structure of nylon 6. Brandburry et al developed a series of lattice constants. Using these lattice constants, unit lattices can be determined correctly, and the reliability of the unit lattices is high.

In this study, the following lattice constants were used:

$$a = 0.482\,\text{nm}, \quad b = 0.782\,\text{nm},$$

and $\quad c = 1.67\,\text{nm} \quad$ (the molecular chain axis is the c-axis).

Although these constants are basically monoclinic systems, they allow ortho-rhombic approximation. In Fig. 13, arcuate reflections are observed in the "edge" and "end" patterns, while the Debye-Scherrer ring is observed in the "through" pattern. This shows that nylon 6 crystals are aligned to the surface layers of the molded NCH specimen in the inside layers. It is found from the diffraction patterns of the surface layers that nylon 6 crystals are uniaxially- and planar-aligned, that the hydrogen-bonding surface (020) or the zigzag plane (110) of the carbon skeleton is aligned parallel with the surface, and that the molecular chain axes exist randomly on the surface. On the other hand, the diffraction patterns differ in the internal layers: the pattern in the "edge" direction differs from that in the "end" direction. This can be explained by considering that the molecular chain of the nylon 6 is uniaxially aligned to the crystals that are perpendicular to the surface of a molded test specimen or the silicate layers. The following facts support this explanation:

- (002) reflections of $2\theta = 12.3°$ are observed in the x-direction.
- (020) and (110) double reflections of $2\theta = 25°$ are observed in the z-direction in both "edge" and "end" patterns.
- (020) and (110) double Debye-Scherrer rings and strong (002) reflections are not observed in the "through" pattern.

Changes in the alignment of nylon 6 crystals were examined relative to their depth from the surface of a molded test specimen. Figure 17 shows how the intensity of the (002) reflection changes relative to the depth from the surface of a molded test specimen. The scattering intensity in the x-direction was measured by introducing the X-ray beam in the y-direction. As the depth increased, the intensity increased dramatically. It stopped increasing at 0.5 mm, and remained constant until the depth reached 1.2 mm. After the depth increased above 1.2 mm, the intensity suddenly dropped around the center of the molded specimen. The change in the intensity up to a depth of 1.2 mm was approximately consistent with the

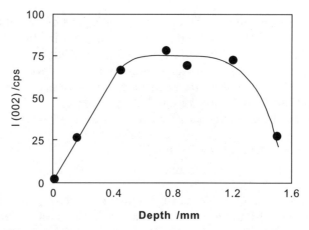

Fig. 17 Peak intensity, I(002), of the 002 reflection of γ-Nylon 6 as a function of depth from the NCH bar surface when the X-ray beam is incident in the y-direction and the scattering intensity was scanned in the x-direction

observations in the X-ray diffraction photographs in Fig. 13. The molecular chain axes of the crystals near the surface of the molded specimen were parallel to the surface. Although they were aligned randomly inside the plane, they changed their bearing toward the direction perpendicular to the surface as the depth increased. Around a depth of 0.5 mm, they were aligned almost perpendicularly. The sudden decrease in the (002) reflections around the center plane was explained as being attributed to the uniaxial alignment of the silicate layers along the flow axis of the resin. Around the center plane of the specimen, the silicate layers were parallel to the flow axis. The molecular chain axes of the nylon 6 crystals that were aligned perpendicular to the silicate layers were aligned randomly around the flow axis. This caused the intensity of the (002) reflection to decrease.

The above results show that NCH consists of three layers: a surface layer, an intermediate layer, and a central layer. Figure 18 shows a schematic representation of this three-layer structure model. In the surface layer, which is located from zero depth (surface) to a depth of 0.5 mm, silicate layers were aligned parallel to the surface, and nylon 6 crystals were uniaxially-aligned along the plane. For example, the (020) or (110) lattice plane was parallel to the plane. On the other hand, the molecular chain axes were aligned randomly inside the plane. In the intermediate layer, from a depth of 0.5 mm

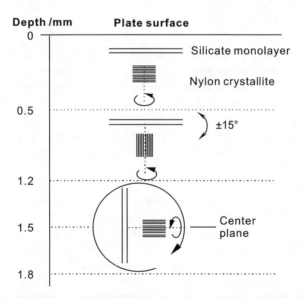

Fig. 18 End-view diagram of the triple-layer structure model for the injection-molded NCH bar 3 mm thick. The flow direction caused by injection-molding is normal to the paper plane. *Curved arrows with one head* mean random orientation round the axis normal to the plane containing the curve. *Arrows with two heads* indicate fluctuation

to a depth of 1.2 mm, the silicate layers were slightly displaced from the direction parallel to the surface. This displacement was within ±15°, which was considered rather large. Nylon 6 crystals were rotated 90° degrees, and aligned almost perpendicular to the surface or the silicate layers. They were aligned randomly around the vertical plane that was perpendicular to the silicate layers. In the center layer, from a depth of 1.2 mm to a depth of 1.8 mm, silicate layers existed in parallel with the flow-axis of the resin. Although the nylon 6 crystals were aligned randomly around the flow axis, the molecular chain axes of each crystal were aligned perpendicular to the silicate layers.

3.8
Other Types of Nylon

After it was verified that nylon 6 could be synthesized with clay to make nanocomposites and to dramatically improve the performance, the same synthesis techniques were applied to other types of nylon resins.

A nylon 66 clay nanocomposite was produced using the dry-compound method [26]. Co-intercalation organophilic clay was used as the clay base. Na-montmorillonite was first processed using hexadecyl trimethyl ammonium ions and epoxy resin. It was then kneaded using a twin screw extruder to make a clay nanocomposite. As the amount of clay that was added increased, the amount of γ (gamma) phases increased. This is thought to be due to the strong interactions between the nylon 66 chains and the surface of the clay layers.

1,10-Diaminodecane and 1,10-decanedicarboxylic acid were polycondensated in the presence of an organophilic clay to polymerize a nylon 1012 clay nanocomposite [27]. X-ray diffraction and TEM observations revealed that the clay layers were exfoliated and uniformly dispersed in nylon 1012. The speed of crystallization of the nanocomposite increased compared with nylon 1012. Furthermore, the tensile strength and the elastic modulus in tension were improved, and the amount of absorbed water was decreased through the improvement of the barrier characteristics.

A nylon 11 clay nanocomposite was prepared using the dry compounding method [28]. X-ray diffraction and TEM observations showed that this technique formed an exfoliated nanocomposite at low concentrations of clay mineral (less than 4 wt %) and that a mixture of exfoliated nanocomposites and interlayer nanocomposites was formed at high concentrations of clay mineral. TGA, DMA and tensile tests showed that the thermal stability and mechanical properties of the exfoliated nanocomposite were superior to those of the interlayer nanocomposite material (with higher clay content). The superior thermal stability and mechanical properties of the exfoliated nanocomposite were thought to be attributed to the fact that the organophilic clay is dispersed stably and densely in the nylon 11 matrix.

Using a clay organized with 12-aminododecanoic acid (ADA), nylon 12 was mixed and polymerized with monomer ADA [29].

3.9
Functions of NCH

3.9.1
Flame Resistance

It is reported that the nylon 6 clay nanocomposite has flame-resistant properties (flammability property). It is thought that a protective layer forms on the surface of this composite and functions to protect the composite from heat. The analysis of this protective layer revealed that it contains an organophilic layer consisting of about 80% clay and 20% graphite [30, 31].

3.9.2
Self-passivation

If the nylon 6 clay nanocomposite is processed in an oxygen plasma, a uniform passivation film is formed. It was found that as the polymers are oxidized, highly oblique composites form, in which the clay concentration increases toward their surfaces, and that the clay layers in these composites function as polymer-protective layers. This indicates that the uniform passivation film may prevent the deterioration of the polymers [32].

4
Polyolefin Clay Nanocomposites

4.1
Introduction

Polyolefin materials (typical polypropylene derivatives) are the type of resins most widely used in the automotive industry. There is a strong need to improve their mechanical properties. After nylon 6 was successfully developed, various research efforts were made to reinforce polyolefins by using clay nanocomposites, but no successful examples of reinforcement using clay nanocomposites was ever reported. It was found for the first time in 1997 that polymers can be intercalated into the clay gallery by using a polyolefin oligomer incorporating hydroxyl groups [15].

In this study, montmorillonite (2C18-Mt) ion-exchanged with dioctadecyl dimethyl ammonium ions and polyolefin with hydroxyl groups on both ends (POLYTEL H made by Mitsubishi Chemical Corporation) were used.

10 g of POLYTEL H was dissolved in 100 ml of toluene. 10 g of 2C18-Mt was added to this solution, which was then stirred strongly to achieve a uniform dispersion of the contents. After this solution was stirred for 10 minutes, the toluene was distilled using an evaporator by placing the solution in an 80 °C water bath. The organophilic montmorillonite obtained this way was called PT-Mt. By changing the ratio of POLYTEL H to 2C18-Mt, the way in which the POLYTEL H affects the swelling behavior of the montmorillonite was examined. The state of swelling was verified by performing X-ray diffraction measurements on various specimens and calculating the interlayer distance.

Figure 19 shows the X-ray diffraction patterns of the 2C18-Mt/POLYTEL H composites mixed in various proportions. When POLYTEL H was added in amounts that were greater than three times the amount of 2C18-Mt, the peaks disappeared completely. It was found from this that POLYTEL H is intercalated between the layers of 2C18-Mt.

Figure 20 shows TEM photographs of sodium-type montmorillonite, 2C18-Mt, and a composite prepared by kneading PT-Mt (one part 2C18-Mt to one part POLYTEL H) and polypropylene. The dispersion of the inorganic matter is on the order of microns, and the dispersibility of the 2C18-Mt increased to the submicron order. Additional POLYTEL H was added so that silicate layers could be dispersed in the PP, which is a non-polar material, and a hybrid material was successfully created.

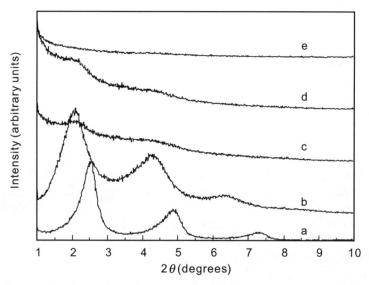

Fig. 19a–e X-ray diffraction patterns for mixtures of 2C18-Mt and polyolefin diol(POLYTEL H): **(a)** 2C18-Mt **(b)** POLYTEL H/2C18-Mt = 1 **(c)** POLYTEL H/2C18-Mt = 3 **(d)** POLYTEL H/2C18-Mt = 5 **(e)** POLYTEL H/2C18-Mt

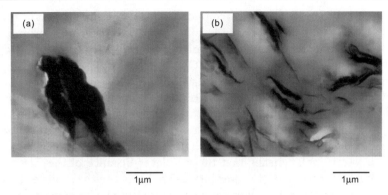

1μm   1μm

Fig. 20a,b TEM images: (**a**) polypropylene/Na-Mt composite, (**b**) polypropylene/2C18-Mt composite

4.2
Fabricating Modified Polypropylene-Clay Nanocomposites

It was found that a functional group must be introduced to intercalate the propylene with no polar group between the organized clay layers [33]. To introduce a functional group to the polymer chain of a polyolefin (including polypropylene) a method of inducing chemical changes using an extruder is applicable. Maleic anhydride-modified polyolefins are produced commercially using such a technique. This type of polyolefin is produced by mixing and melting polyolefin, maleic anhydride and a radical initiator, and then grafting the maleic anhydride group onto the polyolefin.

It has been reported that maleic anhydride-modified polyolefin (a polypropylene with a functional group) has been compounded with organophilic clay by melting and mixing, and the dispersed state of the silicates in the modified polypropylene matrix was investigated [34].

In this study, the following three compounds (each with a different amount of denatured maleic anhydride and different molecular weights) were used as maleic anhydride-modified polyolefins:

- U1001 (Sanyo Chemicals, amount of maleic anhydride: 2.3 wt % Mw: 40 000)
- U1010 (Sanyo Chemicals, amount of maleic anhydride: 4.5 wt % Mw: 30 000)
- PO1015 (Exxon Chemicals, amount of maleic anhydride: 0.2 wt %, Mw: 209 000)

Montmorillonite ion-exchanged with ammonium ions was used as the organophilic clay. This type of montmorillonite is called "C18-Mt".

A nanocomposite made using modified polypropylene and C18-Mt is called a Polypropylene-Clay Nanocomposite (PPCN). Because clay min-

eral is not dispersed in a nanocomposite made by mixing polypropylene
and C18-Mt, this nanocomposite is called a Polypropylene-Clay Composite
(PPCC). A nanocomposite made by mixing polypropylene and talc is called
a Polypropylene-Talc Composite (PPTC).

Figure 21 (a to c) shows X-ray diffraction patterns of the specimens into
which C18-Mt was filled using U1010. The reflection (001) peak associated
with the original layer structure of C18-Mt disappeared, which seemed to be
unrelated to the ratio of the amount of U1010 to that of C18-Mt. However,
a broader peak was observed on the lower-angle side. A shift in the peaks
towards lower angles means that the interlayer distance of the C18-Mt increased; in other words U1010 was intercalated between the layers of C18-Mt.
The position of the highest peak moved toward the lower-angle side as the

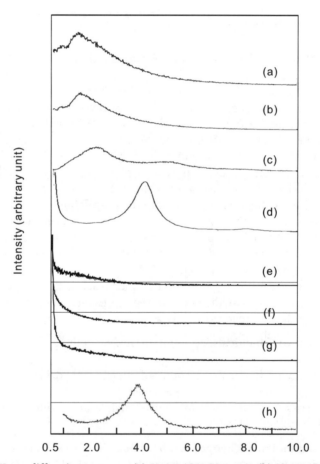

Fig. 21a–h X-ray diffraction patterns: (**a**) U1010/C18-Mt = 3/1 (**b**) U1010/C18-Mt = 2/1
(**c**) U1010/C18-Mt = 1/1 (**d**) C18-Mt (**e**) PPCN(U1010+C18-Mt) (**f**) PPCN(U1001+C18-Mt)
(**g**) PPCN(PO1015+C18-Mt) (**h**) PPCC

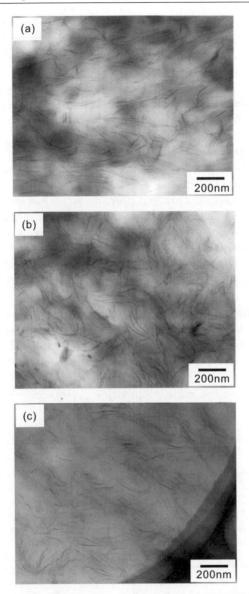

Fig. 22a–c TEM images: (**a**) PPCN(U1010+C18-Mt) (**b**) PPCN(U1001+C18-Mt) (**c**) PPCN (PO1015+C18-Mt)

ratio of U1010 increased. For U1010/C18-Mt 1/1, the interlayer distance was 3.4 nm. For U1010/C18-Mt 2/1, it was 5.7 nm, and for U1010/C18-Mt 3/1, it was 6.3 nm.

Figure 21 (e to g) shows X-ray diffraction patterns of the specimens to which 5 wt% of C18-Mt were added using U1010, U1001 and PO1015. In these X-ray diffraction patterns, no clear peaks are observed in the range of $2\theta = 0.5-10°$. This shows that regularly-layered silicates do not exist, and that the silicate layers are exfoliated. In the X-ray diffraction pattern (Fig. 21h) of PPCC fabricated using polypropylene and C18-Mt, peaks are clearly visible. The interlayer distance is the same as that of the original C18-Mt, indicating that polypropylene is not intercalated between the layers of C18-Mt. Figure 22 shows the TEM images of PPCN. In these photographs, the black lines indicate the cross-sections of the silicate layers, and the gray portions show the modified polypropylene. For all of the specimens, the silicates of C18-Mt were exfoliated, some fine layers were formed, and they were uniformly-dispersed at the nanometer level.

Based on these results, modified polypropylene was intercalated between the organophilic clay layers by melting and kneading organophilic clay and modified polypropylene to which maleic anhydride was introduced as a functional group. It was found that if the amount of added clay was less than 5 wt %, then the silicates were exfoliated into layers and the layers were uniformly dispersed at the nanometer level. Figure 23 shows a schematic representation of the dispersed state of the organophilic clay in the maleic anhydride-modified polyolefin. Because polypropylene with no functional group cannot be intercalated between the layers of organophilic clay, it was thought that the driving force by which the maleic anhydride-modified polyolefin was intercalated between the layers of organophilic clay was generated through an electrostatic interaction between the ammonium ions and the electrical charges on the maleic anhydride groups and the silicates. It was

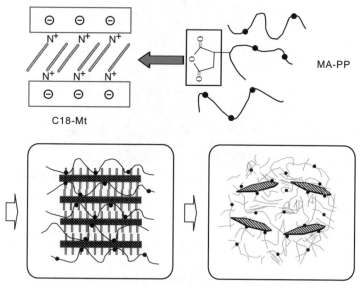

Fig. 23 Schematic representations showing silicate dispersed in modified polyolefin

thought that as the ratio of modified polypropylene increased, more modified polypropylenes were intercalated between the layers of organophilic clay and, as a result, exfoliation occurred.

4.3
Physical Properties of the Modified Polypropylene-Clay Nanocomposites

According to some reports on studies conducted to improve the physical properties of polymer-clay nanocomposites, a small amount of clay was added to a nylon 6-clay nanocomposite and various improvements were achieved: higher polymer strength, higher heat resistance, low linear expansion, low gas permeability, and so on.

This section describes the mechanical characteristics, dynamic viscoelasticity characteristics, and gas permeability characteristics of a modified polypropylene-clay nanocomposite [35]. C18-Mt was used as the organophilic clay, and PO1015 was used as the maleic anhydride-modified polypropylene.

Table 9 shows the results of tensile tests and Izod impact tests. Figure 24 shows the relationship between the elastic modulus in tension and the amount of clay at yield strength. As the amount of added clay was increased, the elastic modulus increased. When 5.3 wt % of inorganic clay was added, it increased by twice as much. The yield strength also increased as the amount of inorganics in the clay was increased. A tendency whereby the yield strength becomes almost saturated by the addition of 2 wt % of inorganics was observed. When 5.3 wt % of inorganics was added to the clay, the elastic modulus increased by a factor of 1.2.

Table 9 Results of tensile tests and Izod impact tests

Sample	Modulus (MPa)	Strength (MPa)	Elongation (%)		Impact strength (J/m)
			5 mm/min	10 mm/min[b]	
PO1015	429	21.1	> 200	> 1000	130
PPCN-2	578 (1.35)	23.2 (1.10)	> 200	756	120
PPCN-3	639 (1.49)	24.0 (1.14)	> 200	688	–
PPCN-4	707 (1.65)	24.7 (1.17)	23.1	–	–
PPCN-5	797 (1.86)	24.9 (1.18)	10.5	–	88
PPTC-2	489 (1.13)	22.5 (1.07)	> 200	–	–
PPTC-5	546 (1.27)	22.9 (1.09)	> 200	–	–
MA2	780	32.5	> 200	–	–
PPCC	830 (1.06)	31.9 (0.98)	105	–	–

[a] The values in parentheses are the relative values of the composites to those of the matrix polymers, respectively
[b] Head speed

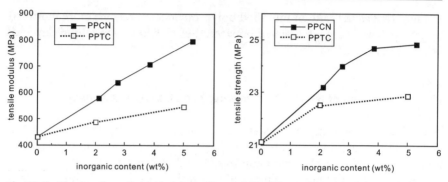

Fig. 24a,b Results of tensile tests: (**a**) tensile modulus (**b**) tensile strength

The elastic modulus and the strength of the PPCN increased markedly compared with PPTC in which talc was dispersed on a sub-micron order. When 4.4 wt % of inorganics were added to the clay, the elastic modulus of a PPCC in which most of the dispersed clays were larger than sub-micron in size was 1.06 times as much as that of neat PP(MA2). The yield strength of this PPCC decreased below that of MA2. The reinforcement effect induced by PPCN in which the dispersed clay silicates are nanosized was much greater than that by PPTC and PPCC, in which the dispersed particulates have sub-micron sizes. This can be explained by noting that a dispersion of nanosized particles allows the area of the interface with the matrix polymer to increase greatly, and therefore the polymer chains can be constrained more strongly. On the other hand, the elongation of the PPCN decreased as the amount of added clay was increased. If the amount of added clay was less than 3 wt %, the elongation was more than 200%, and ductile behavior was observed. At 4 wt %, the elongation decreased to 23%. At 5 wt %, the elongation was 10%, and brittle fracture occurred without the manifestation of a yield point. In the case of PPCN-2, with a small amount of clay, the Izod impact value decreased by about 10%. In the case of PPCN-5, it decreased to about two-thirds. As more clay was added, the impact strength decreased.

Figure 25a shows how the storage modulus of PPCN changes in relation to temperature. Figure 25b shows the relative values of the storage modulus of P01015, which is a matrix polymer. The storage modulus of PPCN was larger than that of P01015 over the whole range of temperature measurement (– 50–130 °C), and it increased as the amount of clay increased. The relative values of the storage modulus of PPCN compared with P01015 increased markedly: they were larger than the glass transition temperature values of PP, and reached a maximum value around 60 °C. The storage modulus of PPCH-5 was 1.5 times as large as that of P01015 at – 50 °C, 2.0 times as large at 30 °C, 2.3 times as large at 60 °C, and 2.0 times as large at 100 °C. The glass transition temperature did not change, even if more clay was added.

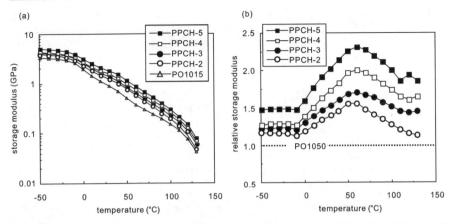

Fig. 25a,b Results of dynamic viscoelastic measurements of PPCN and PO1015: (**a**) storage modulus (**b**) relative storage modulus of PPCN to PO1015

Figure 26 shows the storage modulus and relative storage modulus of PPTC-5, which contains talc. With PPTC-5, the reinforcement effect increased at temperatures higher than the glass transition temperature. However, it only increased by 1.5 times as much, a small increase compared with PPCN. In addition, PPTC-5 did not exhibit a maximum value when the temperature was increased. Figure 27 shows the storage modulus of PPCC and the relative storage modulus of PPCC compared with MA2. Although the reinforcement effect of PPCC increased at a temperature higher than the glass transition temperature, the increase in the reinforcement effect was small: 1.2 to 1.3 times. PPCC also did not exhibit a maximum value when the temperature was increased.

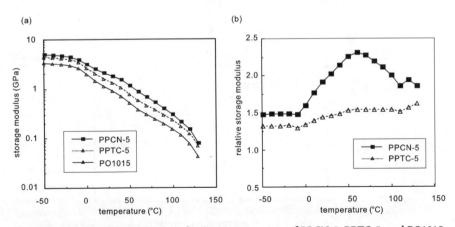

Fig. 26a,b Results of dynamic viscoelastic measurements of PPCN-5, PPTC-5, and PO1015: (**a**) storage modulus (**b**) relative storage modulus of PPCN-5 and PPTC-t to PO1055

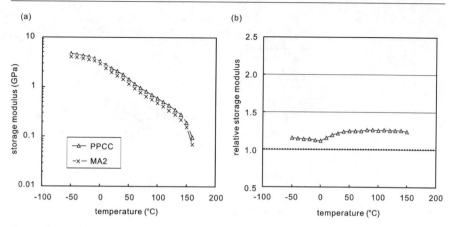

Fig. 27a,b Results of dynamic viscoelastic measurements of PPCC and MA2: (**a**) storage modulus (**b**) relative storage modulus

Figure 28 shows the relationships between the amount of inorganics in the clay and the gas permeability coefficient. The gas permeability coefficient decreased as the amount of added clay increased. The gas barrier performance of PPCN-5 increased by 1.7 times. It has been reported that the barrier performance of the nylon-clay nanocomposites and polymer-clay nanocomposites was improved. This barrier effect is explained as being attributed to the geometrical detour effect of the dispersed nanosized silicates. The barrier effect of PPCN, however, was smaller than that of the nylon-clay nanocomposites. In the case of the nylon-clay hybrid, the addition of 1.8 wt % of montmorillonite caused its hydrogen permeability to decrease to 70%. In the case of PPCN, about 3 wt % of montmorillonite must be added to obtain the same

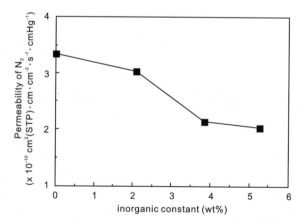

Fig. 28 Gas permeability

effect. Silicate layers are dispersed as monolayers in nylon-clay nanocomposites. In PPCN, the silicate layers exist as double- to quadruple-layer structures. This is thought to have decreased the barrier effect of PPCN.

4.4
Fabricating a Polypropylene-Clay Nanocomposite Using Maleic Anhydride- Modified Polypropylene as a Compatibilizer and Evaluating the Characteristics

To uniformly disperse an organophilic clay with modified polypropylene intercalated as more minute particles, the exfoliation of the silicate layers must be facilitated by adding polypropylene for intercalation, and effective intercalation requires that the modified polypropylene polymers bond smoothly and securely with the polypropylene polymers.

The two factors shown below affect the physical properties of a polypropylene clay nanocomposite made by compounding polypropylene and organophilic clay using modified polypropylene as a compatibilizer [16, 36]. They also affect the dispersibility of silicate layers in this nanocomposite:

1. Compatibility between the polypropylene and the modified polypropylene
2. The type of clay (montmorillonite and synthetic mica)

Three types of maleic anhydride modified polypropylene were used as compatibilizers.

U1010 and U1001 were used as maleic anhydride modified polypropylenes, and MA2 (Japan Polychem Corp.) was used as the polypropylene. Montmorillonite (C18-Mt) exchanged with octadecylammonium ions and synthetic mica (C18-Mc) were used as the clays.

4.4.1
Effects of the Compatibility between Modified Polypropylene and Polypropylene

It is thought that the compatibility between polypropylene and maleic anhydride-modified polypropylene is affected greatly by the amount of maleic anhydride groups in the modified polypropylene. A polypropylene clay nanocomposite was fabricated using two types of maleic anhydride-modified polypropylene (each with different numbers of maleic anhydride groups) as compatibilizers. The physical properties of this nanocomposite and the dispersibility of the silicate layers within it were examined. The ways in which the method used to mix the polypropylene, the modified polypropylene and the organophilic clay affect the resulting physical properties were also examined. The results are summarized in the following subsections.

4.4.1.1
Compatibility Between Modified Polypropylene and Polypropylene

Figure 29 shows photographs of MA2/U1010 and MA2/U1001 in a molten state at 200 °C observed under an optical microscope. With MA2/U1010, non-uniformity was observed, and macroscopic phase-separation was noted. With MA2/U1001, no non-uniformity was observed using visual or optical means, indicating that the compatibility is better in MA2/U1001 than in MA2/U1010. U1010 contains 4.5 wt % of maleic anhydride groups, while U1001 contains only 2.3 wt %. Therefore, the compatibility with polypropylene is higher in U1001.

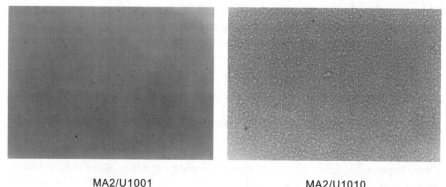

MA2/U1001   MA2/U1010

Fig. 29 Miscibility of PP and modified PP

4.4.1.2
Dispersibility of Silicate Layers

Figure 30d and e show the X-ray diffraction patterns of a nanocomposite after being compounded with MA2. In the case of the U1010-type PPCN (PP+U1010/C18-Mt), which has low compatibility with MA2, the peak associated with the layered clay structure in the U1010/C18-Mt interlayer compound was clearly observed around $2\theta = 0.7°$ after being mixed with MA2. In the case of the U1001-type PPCN, which has high compatibility with MA2, the peaks observed with the U1001/C18-Mt interlayer compound at around $2\theta = 0.8°$ were not observed; instead, gently-sloping shoulder lines were observed around $2\theta = 1.5°$. In the case of the U1001-type PPCN (PP+U1001/C18-Mt), the regularity of the layered structure is thought to have decreased, although the silicate layers still remain.

Figure 31a and b show the TEM images. Exfoliated silicate layers are dispersed more widely and uniformly in the U1001-type PPCN (PP+U1001/C18-Mt) than in the U1010-type PPCN (PP+U1010/C18-Mt). Many silicate layers, whose interlayer distances were expanded to more than 5 nm, were also

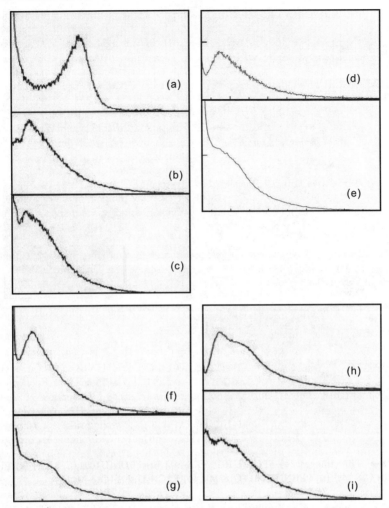

Fig. 30a–i X-ray diffraction patterns: (**a**) C18-Mt (**b**) U1010/C18-Mt (**c**) U1001/C18-Mt (**d**) PPCN(U1010/C18-Mt) (**e**) PPCN(U1001/C18-Mt) (**f**) PPCN(U1010+C18-Mt) (**g**) PPCN(U1001+C18-Mt) (**h**) PPCN(U1010+C18-Mc) (**i**) PPCN(U1001+C18-Mc)

observed (as shown in the X-ray diffraction patterns), indicating that the dispersibility of the silicate layers in the U1001-type PPCN is better than that in the U1010-type PPCN.

Figures 30f,g,h and i show the X-ray diffraction patterns of the other specimens that were fabricated. In these X-ray diffraction patterns, the peak around $2\theta = 4°$ associated with the layered structure of organophilic clay (C18-Mt, C18-Mc) was not observed. It was noted that the pattern is rising toward the low-angle side. This shows that polymer chains were intercalated

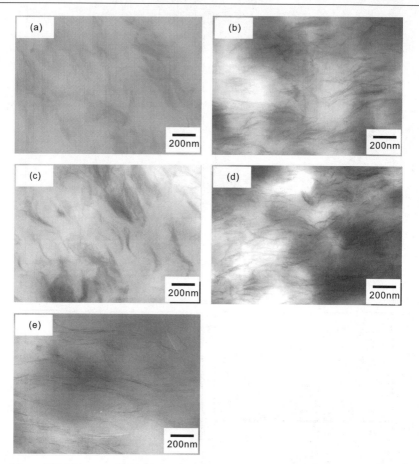

Fig. 31a–e TEM images: (**a**) PPCN(U1010/C18-Mt) (**b**) PPCN(U1001/C18-Mt) (**c**) PPCN (U1010+C18-Mt) (**d**) PPCN(U1001+C18-Mt) (**e**) PPCN(U1001+C18-Mc)

between the layers of organophilic clay during melting and kneading and, as a result, the interlayer distance expanded.

With U1010-type C18-Mt and C18-Mc, a peak associated with the layered structure of the silicate was observed around $2\theta = 0.8°$. It is inferred from this that silicate layers with U1010 intercalated are dispersed and that they maintain their layered structure. The interlayer distance for U1010-type PPCN (PP+U1010+C18-Mt) (Fig. 30f) is equivalent to that of the U1010/C18-Mt interlayer compound (Fig. 30b) which was fabricated using U1010 and C18-Mt in the same proportions. Therefore, it is thought that U1010 is selectively-intercalated between the layers of C18-Mt during mixing, even if polypropylene is mixed simultaneously.

With the U1001-type C18-Mt, the clear peak around $2\theta = 0.8°$ associated with the layered structure of clay was not observed. The layered structure was less regular than that in the U1010-type nanocomposites. The same result was obtained with the specimens fabricated using the U1001/C18-Mt interlayer compound. In the case of C18-Mc, the peak associated with the layered structure of the silicate was observed on the low-angle side at angles lower than $2\theta = 0.5°$.

The nanocomposites fabricated using mica are described in detail in Sect. 4.4.2

Figures 31c,d and e show the TEM images. With the U1001-type C18-Mt and C18-Mc, the dispersibility of the silicate layers was better than that in the U1010-type PPCN (PP+U1010+C18-Mt). The dispersibility of the silicate layers in specimens fabricated using a simultaneous mixing method was equivalent to that in specimens fabricated using clay interlayer compounds.

The use of U1001, which is highly compatible with MA2, allowed the interlayer distance in the silicate layers to expand, the layer exfoliation process to be facilitated, and fine silicate particles to be dispersed in MA2, whichever fabrication method was used. It was verified from this that the compatibility between the polypropylene and the modified polypropylene greatly affects the dispersibility of the clay. It was also verified that if the same percentage composition is used to fabricate nanocomposites, the resultant dispersibility of the silicate layers is equivalent among the fabricated nanocomposites.

4.4.1.3
Physical Properties of Nanocomposites

Figure 32 shows the temperature dependence of the storage moduli, calculated from the dynamic viscoelasticity measurements of the U1010-type PPCN (PP+U1010+C18-Mt), the PPCN (PP+U1010+C18-Mc), the U1001-type PPCN (PP+U1001+C18-Mt), and the PPCN (PP+U1001+C18-Mc), all of which were fabricated using a simultaneous mixing method. Figures 33a and b show the relative storage moduli of each single MA2 in the individual specimens, which were calculated based on the data in Fig. 32. Table 10 shows the storage moduli at the main temperatures and the glass transition temperatures (T_g) calculated from the $\tan \delta$ peaks.

In the temperature zone of 50 °C or higher, the relative modulus changed due to the effects of the type of the modified polypropylene, and U1001 exhibited a higher storage modulus than U1010. The same tendency was noted with C18-Mt and C18-Mc. This was thought to be attributed to the facts that the melting point (softening point) of U1001 is higher, and that U1001 has superior dispersibility.

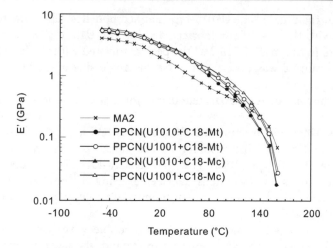

Fig. 32 Storage modulus of PPCN and MA2

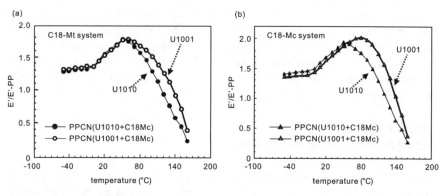

Fig. 33 Relative storage modulus of the PPCN using C18-Mt and C18-Mc

4.4.2
Effect of the Type of Clay

This section describes how the dispersed state of the silicate layers in a polypropylene clay nanocomposite and the physical properties of this nanocomposite are affected by the type of clay used. Two types of clay were used in this experiment: organophilic montmorillonite and organophilic mica.

4.4.2.1
Dispersibility of Clay

For the U1010-type C18-Mt and C18-Mc specimens fabricated using a simultaneous mixing method, clear peaks were observed around $2\theta = 0.8°$

(interlayer distance: about 6 nm) in the X-ray diffraction patterns, as shown in Fig. 30f and Fig. 30h. As shown in the TEM image of the PPCN (PP+U1010+C18-Mt) (Fig. 31c), it is believed that the silicates were dispersed and layered. In the case of the U1001-type PPCN (PP+U1001+C18-Mc) using C18-Mc (Fig. 30i), the peak associated with an interlayer distance of 11.4 nm was observed. With the U1001-type PPCN (PP+U1001+C18-Mt) using C18-Mt (Fig. 30g), shoulder lines ($2\theta = 2°$ near field, interlayer distance: 4 to 5 nm) that continue to rise toward the low-angle side were observed. It was inferred from this that, although the layered structure of silicate remains in both C18-Mt and C18-Mc specimens, the interlayer distance of the C18-Mc expanded by more than that of the C18-Mt, because shoulder lines were observed around $2\theta = 2°$ in C18-Mt.

Figures 31d and e show the TEM images. For the PPCN (PP+U1001+C18-Mt) using montmorillonite, the interlayer distance expanded by more than several nanometers, as shown in the X-ray diffraction patterns, and the silicate layers were uniformly dispersed in single- or multiple-layers. In the case of the PPCN (PP+U1001+C18-Mc) using mica, an interlayer distance of about 10 nm was maintained, and silicate layers were likewise uniformly dispersed in monolayers or multilayers. The sizes of the silicate layers, however, were different: the size of the silicate layers in the PPCN (PP+U1001+C18-Mc) using mica was larger than that of silicate layers in the PPCN using montmorillonite. This is thought to be due to the fact that the original size of the silicate of the synthetic mica was larger than the size of the silicate in montmorillonite.

4.4.2.2
Physical Properties

Figures 33a and b show the relative storage moduli of the PPCN using montmorillonite and the PPCN using mica. The relative storage moduli of both the U1010 and U1001 specimens using mica were high over the whole temperature range.

The relative storage moduli of the U1010 specimens reached maxima at 50 °C and their temperature dependence patterns were similar, whichever type of clay was used. The temperature dependence patterns of the relative storage moduli of the U1001 specimens varied depending on the type of clay used.

All of the specimens exhibited the same or similar relative storage moduli at temperatures below T_g (10 °C). In the temperature range above T_g up to around 100 °C, specimens using mica exhibited a higher reinforcement effect. The relative storage modulus of the PPCN (PP+1001C8-Mt) using mica was twice that of MA2 at 80 °C.

4.5
Fabricating a Polyethylene-Clay Nanocomposite and Evaluating its Physical Properties

A polyethylene nanocomposite was fabricated using the following materials:

- Polyethylene (KF380, melt flow index (MFR) 4.0 g/10 min ASTM D1238 (190 °C, 2.16 kg); PE) supplied by Japan Polychem Corp.
- Maleic anhydride-modified polyethylene (Fusabond 226D, base resin type LLDPE, grafted maleic anhydride 0.90 wt %, MFR 1.5 g/10 min ASTM D1238 (190 °C, 2.16 kg); MA-g-PE) supplied by E. I. DuPont

C18-Mt was used as the organophilic clay. Table 11 shows the composition of this nanocomposite. Figure 34 shows the X-ray diffraction patterns of the materials. The clay was uniformly dispersed, as shown. Tables 12 and 13 show the mechanical characteristics and the nitrogen gas permeability characteristics, respectively [37]. Materials with high rigidity and high gas barrier characteristics were obtained, as in the case of polypropylene.

Table 10 Dynamic viscoelastic measurement results

| | Storage modulus (GPa) | | | |
	-40 °C	20 (°C)	80 (°C)	140 (°C)
PPCN (U1010/C18-Mt:3/1)	5.06 (1.29)	2.98 (1.51)	1.14 (1.76)	0.202 (0.91)
PPCN (U1010/C18-Mt:2/1)	4.90 (1.25)	2.85 (1.44)	1.05 (1.62)	0.237 (1.07)
PPCN (U1010/C18-Mt:1/1)	4.50 (1.15)	2.57 (1.30)	0.887 (1.37)	0.246 (1.11)
PPCC	4.50 (1.15)	2.36 (1.19)	0.010 (1.26)	0.278 (1.25)
U1010/MA2 : 22/78	3.92 (1.00)	1.99 (1.01)	0.597 (0.92)	0.153 (0.69)
U1010/MA2 : 7/93	3.80 (0.97)	1.97 (0.99)	0.612 (0.94)	0.19 (0.86)
MA2	3.92	1.98	0.648	0.222

[a] The values in parentheses are the relative values of the composites to those of the matrix MA2, respectively

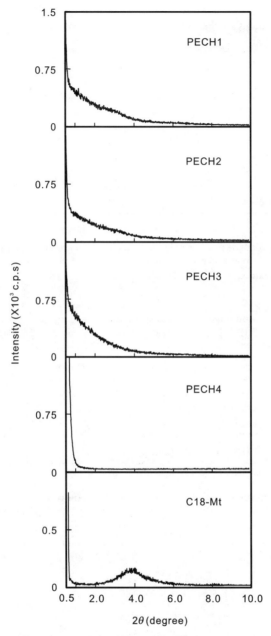

Fig. 34 X-ray diffraction patterns of the polyethylene-clay nanocomposites, a related sample (note that PECH4 is the same as PE1 in Table 11), and C18-Mt

Table 11 Compositions of the prepared nanocomposites based on PE, MA-g-PE, and C18-Mt

Sample name	Clay	Composition (weight ratio)		Inorganic content (%)
		PE	Ma-g-PE	
PE	–	100	0	0
MA-g-PE	–	0	100	0
PE1	–	70	30	0
PECH1	C18-Mt	70	30	5.4
PECH2	C18-Mt	70	30	3.5
PECH3	C18-Mt	0	100	5.2
PECC	Na-Mt	70	30	5.7

Table 12 Tensile properties of the polyethylene-clay nanocomposite and related samples ($n = 5$). The values in parentheses are the relative values of the nanocomposites and PECC to those of each matrix.

Sample name	Tensile properties			
	Modulus (MPa)	Yield strength (MPa)	Yield strain (MPa)	Ultimate elongation (%)
PE	102	7.3	7.1	> 500
MA-g-PE	118	9.3	8.0	180
PE1	99	7.5	7.7	> 500
PECH1	180 (1.8)	10.3 (1.4)	5.6	> 500
PECH2	140 (1.4)	9.4 (1.3)	6.8	> 500
PECH3	157 (1.3)	12.6 (1.4)	7.0	155
PECC	103 (1.0)	7.9 (1.1)	8.4	> 500

Table 13 Gas permeabilities of the polyethylene clay nanocomposites and related samples($n=2$)

Sample name	Gas permeability coefficient $\times 10^{13}$ $(cm^3 \, (STP) \, cm \, cm^{-2} s^{-1} Pa^{-1})$
PE	5.26
MA-g-PE	5.46
PE1	5.32
PECH1	3.78
PECH2	3.91
PECH3	3.48
PECC	5.48

4.6
Fabricating an Ethylene Propylene Rubber-Clay Nanocomposite and Evaluating the Characteristics

Ethylene propylene rubber (EPR) was used to develop a new type of olefin clay nanocomposite [38].

MP0610, supplied by Mitsui Chemicals, was used as the maleic anhydride-modified EPR. The concentration of maleic anhydride groups was 0.42 wt % (4.81 mg KOH/g). The molecular weight obtained by GPC measurement was

$$M_n : 125\,000 \quad \text{and} \quad M_w : 397\,000 \tag{13}$$

Figure 35 shows the X-ray diffraction patterns of EPR-CN and C18-Mt. The diffraction peak shown in this figure is associated with the reflection on the surface (001) of layered silicates. No clear diffraction peak was observed in the X-ray diffraction pattern of EPR-CN. This shows that regularly-layered silicates do not exist in EPR-CN. Figure 36 shows the TEM image of EPR-CN6. The silicates were exfoliated and uniformly-dispersed at the nanometer level. The use of a small concentration of maleic anhydride groups caused the silicates to become uniformly dispersed at the nanometer level in the EPR, as in the case of modified polypropylene.

Figure 37a shows the S-S curve obtained by conducting a typical tensile test on the specimens. MP0610 exhibited a yield point around the 80% elongation point (as an elastomer normally does) and fractured around the 900% elongation point. With EPR-CN, it became difficult to identify the yield point as the amount of clay increased, and the elongation rate also decreased.

In particular, EPR-CN8 fractured without manifesting a yield point. The elastic modulus of EPR-CN increased as the amount of clay increased. It was

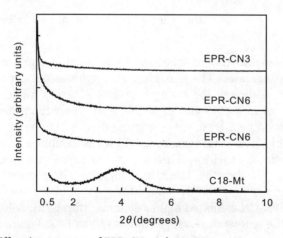

Fig. 35 X-ray diffraction patterns of EPR-CN, and C18-Mt

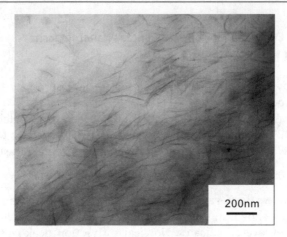

Fig. 36 TEM image of EPR-CN6

(a)   (b)

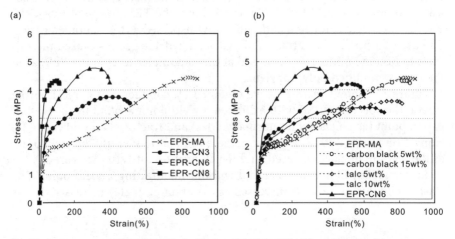

Fig. 37a,b Representative stress-strain curves: (**a**) EPR-CNs (**b**) conventional composites

three times as large as that of MP0610. The elongation rate decreased as the amount of clay increased. The maximum strength was not dependent on the amount of clay. Although the strength of EPR-CN6 was higher than that of MP0610, those of EPR-CN3 and EPR-CN8 were lower than that of MP0610. Figure 37b shows the S-S curves of commonly-used composite materials to which carbon black and talc were added. EPR-CN exhibited a very large elastic modulus, compared with specimens with the same amount of additive or dopant material. On the other hand, its elongation rate decreased.

Figure 38 shows how the dynamic storage modulus changes relative to temperature. EPR-CN exhibited higher dynamic storage modulus values than MP0610 over the temperature range from – 150 to 1 °C. Figure 39 shows the dynamic storage modulus of EPR-CN relative to the dynamic storage modu-

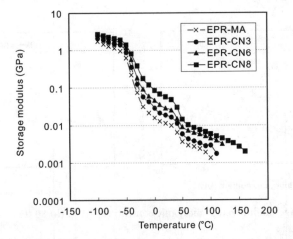

Fig. 38 Storage moduli of EPR-CNs

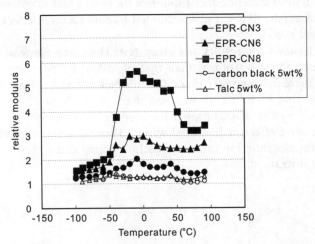

Fig. 39 Relative storage moduli of EPR-CNs and conventional composites

lus of MP0610 using conventional composite materials. The relative storage modulus of EPR-CN was small at temperatures below the glass transition temperature (about $-35\,^\circ$C as calculated from $\tan\delta$). However, it increased dramatically at temperatures above T_g; after it reached a maximum value between $-20\,^\circ$C and $0\,^\circ$C, it began to decrease again.

Figure 40 shows the relationship between the dynamic storage modulus of EPR-CN and the amount of inorganics it contains, relative to the relationship between the dynamic storage modulus of a conventional composite material and the amount of inorganics that it contains. The storage modulus of EPR-CN6 at $20\,^\circ$C is almost equal to that of a composite material with 30 wt % of inorganics (Fig. 40a). Organophilic clay dispersed at the nanometer level

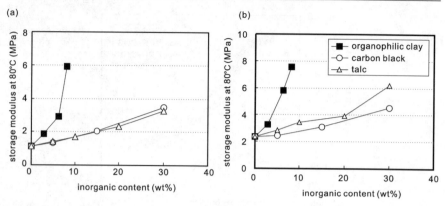

Fig. 40a,b Storage moduli versus inorganic content: (**a**) 20 °C (**b**) 80 °C

exhibited a reinforcement effect about five times as great as conventional re-inforcing materials. EPR-CN also exhibited a reinforcement effect about five times as great at 80°.

Figure 41 shows the results of a creep test. The creep elongation of EPR-CN was much more restrained than that of MP0610. The creep elongation of MP0610 increased by more than 50% in one hour and it fractured in two hours, while that of EPR-CN6 was less than 1% in 30 hours. A composite material with 5 wt % of carbon black added did not exhibit a conspicuous creep-restraining effect; it fractured within 3 hours. It was thought that the dispersed silicates function as large crosslink points and so the creep can be successfully restrained.

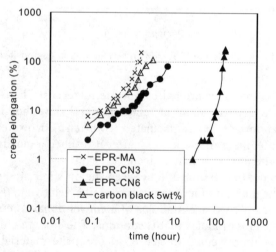

Fig. 41 Creep test results

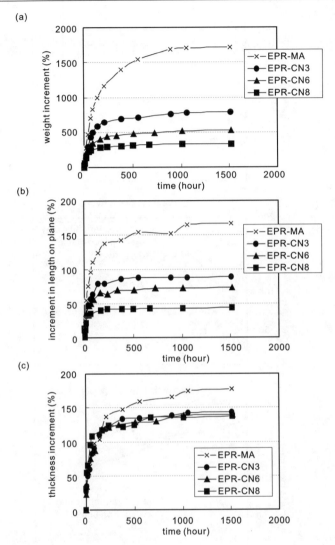

Fig. 42a–c Swelling test results: (**a**) weight increments (**b**) length increments on plane (**c**) thickness increments

Figure 42a and b show how the weights, planar dimensions and thicknesses of the specimens (10 mm × 10 mm × 2 mm) increased when they were immersed in hexadecane at 25 °C. The degree of swelling in EPR-CN was restrained more conspicuously than the degree of swelling in MP0610. While the weight of MP0610 increased by more than 1700%, the increase in weight of the EPR-CN8 was restricted to 333%. The increase of the planar dimensions of EPR-CN was also noticeably restrained as the amount of added clay was increased. The increase in the planar dimensions of MP0610 was 170%,

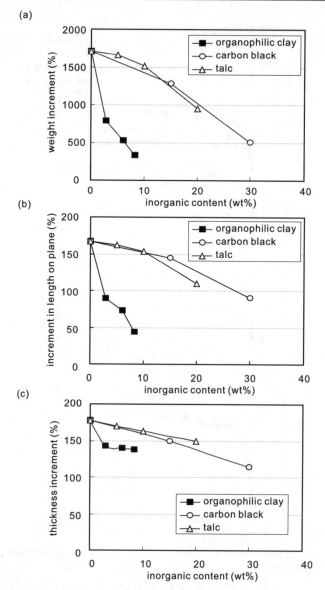

Fig. 43a–c Swelling increment versus inorganic content: (**a**) weight increments (**b**) length increments on plane (**c**) thickness increments

while of the increase for EPR-CN8 was restrained to 45%. On the other hand, increases in the thicknesses of the specimens could scarcely be restrained by adding clay. With EPR-CN, noticeable anisotropic characteristics were observed with respect to the relationship between the increase of the planar dimensions and the increase in thickness. Figure 43 shows relationships be-

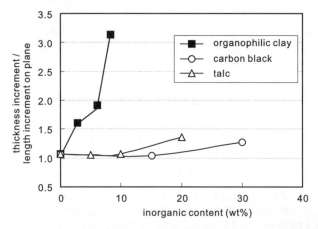

Fig. 44 Anisotropy between length increment and thickness increment versus inorganic content

tween the increase in the degree of swelling and the amount of inorganic content after EPR-CN was immersed for 1500 hours. As is evident from Fig. 43, organophilic clay is far superior to other conventional reinforcing materials with respect to the overall swelling restraining effect, particularly in regard to the weight-restraining and planar dimension-restraining effects. The increase in the weight of EPR-CN3 was equivalent to that of the weight of a composite material to which 20 wt % of a conventional reinforcing material was added. Figure 44 shows the anisotropy between the increase in the planar dimensions and the increase in thickness. The anisotropy of EPR-CN is considerably larger than that of conventional reinforcing materials. Conventional composite materials do not exhibit anisotropy if the amount of added reinforcing material is less than 10 wt %, while EPR-CN8 exhibits an anisotropy of 2.5 or greater. It is known that both the polymer chains and the silicates dispersed at nanometer levels are aligned in parallel with the sheet surface fabricated by compression molding. It was thought that parallel-aligned silicate layers and polymer chains make it possible to selectively-restrain the increase in the planar dimensions in EPR-CN.

4.7
Synthesizing an Ethylene Propylene Diene Rubber (EPDM)-Clay Nanocomposite and Evaluating its Characteristics

Polyolefin must be modified to make it become polarized. This polarized polyolefin can then be processed to synthesize a nanocomposite. Rubber does not need to be modified to synthesize a nanocomposite. If we take EPDM as an example, EPDM can be mixed with C18-Mt, and this mixture is intercalated in the clay gallery during vulcanization. When the vulcanization

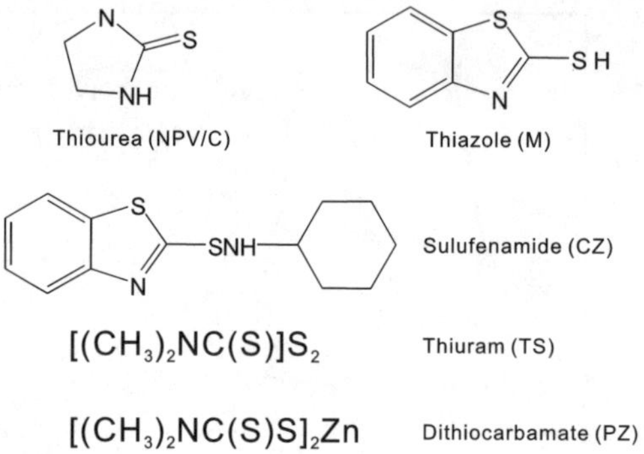

Fig. 45a–e Structual formulae of vulcanization accelerators: (**a**) ethylenethiourea, NPV/C (**b**) 2-mercaptobenzothiazole, M (**c**) N-cyclohexyl-2-benzothiazylsulfenamide, CZ (**d**) tetramethylthiuram monosulfide, TS (**e**) zinc dimethyldithiocarbamate, PZ

accelerator is dissociated during vulcanization, sulfur radicals form and combine with the EPDM molecules. The sulfur radicals combined with EPDM molecules form polar groups, which in turn intrude into the clay gallery spacing. A type of a vulcanization accelerator including disulfide bonds must be used, as shown by (d) and (e) in Fig. 45. As vulcanization and intercalation take place simultaneously, this process is called "in situ intercalation". The principles of this process are shown in Fig. 46 [39].

4.8
Synthesizing a Polyolefin-Clay Nanocomposite Using the Polymerization Method

In 1999, an attempt was made to synthesize a polyolefin in clay gallery. Specifically, clay was ion-exchanged using tetradecylammonium ions, and this ion-exchanged clay and a palladium-based complex of the Brookhart-type were mixed and conditioned in toluene to polymerize ethylene. The interlayer distance was initially 1.99 nm. This increased to 2.76 nm after a palladium catalyst was added. It was confirmed that the X-ray peak disappeared 24 hours after ethylene was introduced [40].

There are reports that ethylene and 1-octane have been copolymerized using a similar catalyst, and that the physical properties of this copolymer were compared with those of a nanocomposite fabricated using the dry-compound method [41]. A case is also reported in which polyethylene was subjected to in-situ polymerization.

Hydroxy-groups were made to react with an aluminum compound (for example, triisobutylaluminum) between clay layers. After this compound was washed and thoroughly dried, it was brought into contact with vinyl alcohol (for example, ω-undecylenylalcohol). The clay made in this way included vinyl groups between the layers. If a polymerization catalyst and ethylene

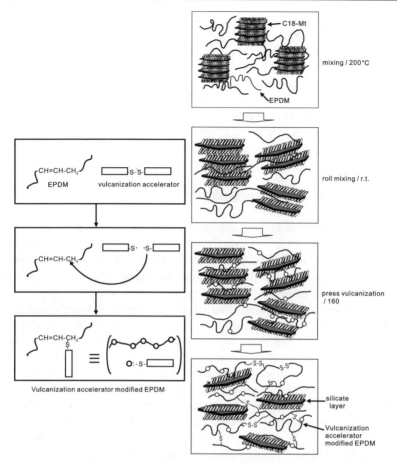

Fig. 46 Schematic diagram of the vulcanization process for EPDM and intercalation into clay gallery

were intercalated into the clay gallery, the vinyl groups would react with ethylene such that the ethylene at the chain end and the clay combined to create a new nanocomposite [42]. If ammonium ions are bonded to the polypropylene, this nanocomposite can be made by dry compounding.

PP (ammonium group-terminated PP) with terminal ammonium groups was synthesized (M_n = 58900 and M_w = 135 500 g/mol; T_m = 158.2 °C) using a Zr catalyst. This PP was dry-compounded with clay (2C18-Mt) conditioned using dioctadecylammonium, and a nanocomposite was obtained.

It has been pointed out that a nanocomposite in which clay is completely dispersed (exfoliated) can be obtained by using the PP with ammonium linked at the chain end instead of using PP with a functional group at the side chain [43].

5
Green Nanocomposites

The bio-related resin polylactic acid is well known as a renewable material. However, renewable materials such as this lack the mechanical and thermal properties to be of any practical use. In order to overcome these drawbacks, the synthesis of clay nanocomposites based on renewable materials has been discussed. These materials are known as 'Green Nanocomposites' [44] and are now forming a new sector in materials studies.

5.1
Bio-related Polymer-Clay Nanocomposites

A typical bio-related polymer, polylactic acid (PLA), has been studied. By mixing C18-Mt and PLA, the interlayer distance of the clay increases and a nanocomposite can be obtained. However, the clay has not yet reached the exfoliated condition [45, 46]. Recently, the authors have successfully developed a complete exfoliated PLA clay nanocomposite using clay (C18(OH)-Mt) that is substituted by bis(2-hydroxyethyl) methyl octadecylammonium via the open cyclic polymerization of lactide from OH groups [47]. For nylon compounds, clay and polymers are interacted by ion-bonding or hydrogen bonding and in the polyolefin compounds they are interacted by modification and hydrogen bonding; however, polyester compounds such as PLA do not interact strongly. Therefore, the mechanical or thermal properties are not significantly improved, although the effects of the clay on the crystallization are significant. Furthermore, a clay nanocomposite was synthesized using a mixture of C18-Mt and polybutylene succinate (PBS) [48]. A chitosan-clay nanocomposite has also been synthesized by using a solvent [49].

5.2
Plant Oil-Clay Nanocomposite

Plant oils are produced from natural materials; however, they have seldom been used for industrial purposes except as edible oils. In order to use plant oils more effectively, their use in clay nanocomposites has been discussed [50].

Triglyceride oils have been extensively used for various applications such as coatings, inks, and agrochemicals. These oil-based polymeric materials, however, do not exhibit the rigidity and strength required for structural applications by themselves. In some cases, therefore, triglyceride has been used as a minor component in polymeric materials; it was used solely as a modifier to improve the physical properties of the polymeric material.

An epoxidized triglyceride oil was subjected to intercalation into an organically-modified clay, followed by acid-catalyzed curing of the epoxy-containing triglyceride, leading to the production of a new class of biodegradable-nanocomposites from inexpensive renewable resources.

C18-Mt and ESO were mixed using a solvent. The solvent was then washed out and cast films were produced. Uniform films are obtained using 5% to 20% of clay. XRD and TEM measurements confirmed that the clay was dispersed and flexible films were successfully produced. These films could be used as a coating medium.

6
Conclusion

Currently, a variety of polymer nanocomposites have been developed and many of them now have practical applications. Clay hybrid materials that exhibit high compensating effects upon the addition of small amounts of additives have attracted attention worldwide, along with the effect of gas barriers, and major chemical manufacturing companies have been involved in research into these materials. The areas where these nanocomposites may be used include:

1. Resin materials for molding; in particular, automotive components that require enhanced hardness characteristics.
2. Use in thin-film materials; in particular, packing films for foods.
3. Use in rubber materials that require barrier performance; in particular, hoses for automotive use.
4. Use in resin components for domestic electrical appliances that require flame resistance.

Expected effects of the widespread interest in these nanocomposites include:

1. Low-weight benefits can be expected due to the high-performance effect of small amounts of additives compared with other fillers (including glass fibers).
2. Easy to recycle because no filler needs to be broken down when reprocessing.

However, there are some anticipated problems with these materials:

1. Reinforcement can be achieved with a small amount of added clay; however, if large amounts were added, the impact resistance could decrease.
2. Synthetic clays that could be superior in function and cost to natural clays have not yet been developed.

In summary:

1. The hybridization of clay can be used for the reinforcement of various resin materials and these can replace glass fiber reinforcement materials.
2. The reinforcement mechanism has yet to be clarified. Therefore, collaboration between industry-government-academia will help to create new materials.

In addition to the nylon and polypropylene that were described here, he hybridization of polyimide [51] has also been achieved, as has polystyrene-clay hybridization [52]. It is hoped that clay hybridization can be used as a standard method for reinforcing resin materials (such as glass fiber).

References

1. Ray SS, Okamoto M (2003) Prog Polym Sci 28:1539
2. Usuki A, Kojima Y, Kawasumi M, Okada A, Kurauchi T, Kamigaito O (1993) J Mater Res 8:1174
3. Usuki A, Kawasumi M, Kojima Y, Fukushima Y, Okada A, Kurauchi T, Kamigaito O (1993) J Mater Res 8:1179
4. Kojima Y, Usuki A, Kawasumi M, Fukushima Y, Okada A, Kurauchi T, Kamigaito O (1993) J Mater Res 8:1185
5. Messersmith PB, Giannelis EP (1995) J Polym Sci A Polym Chem 33:1047
6. Lan T, Pinnavaia TJ (1994) Chem Mater 6:2216
7. Usuki A, Okamoto K, Okada A, Kurauchi T (1995) Kobunshi Ronbunshu 52:728
8. Biasci L, Aglietto M, Ruggeri G, Ciardelli F (1994) Polymer 35:3296
9. Choo D, Jang LW (1996) J Appl Polym Sci 61:1117
10. Moet A, Salahuddin N, Hiltner A, Baer E, Akelah A (1994) Mat Res Soc Symp Proc 351:163
11. Fukumori K, Usuki A, Sato N, Okada A, Kurauchi T (1991) In: Kimpara I (ed) Proc 2nd Japan International SAMPE Symp. Society for the Advancement of Material and Process Engineering, Kamakura, Japan, p 89
12. Kojima Y, Fukumori K, Usuki A, Okada A, Kurauchi T (1993) J Mater Sci Lett 12:889
13. Yano K, Usuki A, Okada A, Kurauchi T (1993) J Polym Sci A Polym Chem 31:2493
14. Lan T, Kaviratna PD, Pinnavaia TJ (1994) Chem Mater 6:573
15. Usuki A, Kato M, Okada A, Kurauchi T (1997) J Appl Polym Sci 63:137
16. Kawasumi M, Hasegawa N, Kato M, Usuki A, Okada A (1997) Macromolecules 30:6333
17. Uribe-Arocha P, Mehler C, Puskas JE, Altstadt V (2003) Polymer 44:2441
18. Kojima Y, Usuki A, Kawasumi M, Okada A, Kurauchi T, Kamigaito O (1993) J Polym Sci A Polym Chem 31:1755
19. Liu L, Qi Z, Zhu X (1999) J Appl Polym Sci 71:1133
20. Cho JW, Paul DR (2001) Polymer 42:1083
21. Fornes TD, Yoon PJ, Keskkula H, Paul DR (2001) Polymer 42:9929
22. Shah RK, Paul DR (2004) Polymer 45:2991
23. Hasegawa N, Okamoto H, Kato M, Usuki A, Sato N (2003) Polymer 44:2933
24. Usuki A, Koiwai A, Kojima Y, Kawasumi M, Okada A, Kurauchi T, Kamigaito O (1995) J Appl Polym Sci 55:119

25. Kojima Y, Usuki A, Kawasumi M, Okada A, Kurauchi T, Kamigaito O, Kaji K (1995) J Polym Sci B Polym Phys 33:1039
26. Liu X, Wu Q, Berglund LA (2002) Polymer 43:4967
27. Wu Z, Zhou C, Qi R, Zhang H (2002) J Appl Polym Sci 83:2403
28. Liu T, Lim KP, Tjiu WC, Pramoda KP, Chen ZK (2003) Polymer 44:3529
29. Kim GM, Lee DH, Hoffmann B, Kressler J, Stoppelmann G (2001) Polymer 42:1095
30. Gilman JW, Jackson CL, Morgan AB, Harris Jr RH, Manias E, Giannelis ER, Wuthenow M, Hilton D, Phillips SH (2000) Chem Mater 12:1866
31. Kashiwagi T, Harris RH Jr, Zhang X, Briber RM, Cipriano BH, Raghavan SR, Awad WH, Shields JR (2004) Polymer 45:881
32. Fong H, Vaia RA, Sanders JH, Lincoln D, Vreugdenhil AJ, Jiu W, Bultman J, Chen C (2001) Chem Mater 13:4123
33. Hasegawa N, Okamoto H, Kawasumi M, Kato M, Tsukigase A, Usuki A (2000) Macromol Mater Eng 280/281:76
34. Kato M, Usuki A, Okada A (1997) J Appl Polym Sci 66:1781
35. Hasegawa N, Okamoto H, Kato M, Usuki A (2000) J Appl Polym Sci 78:1918
36. Hasegawa N, Kawasumi M, Kato M, Usuki A, Okada A (1998) J Appl Polym Sci 67:87
37. Kato M, Okamoto H, Hasegawa N, Tsukigase A, Usuki A (2003) Polym Eng Sci 43:1312
38. Hasegawa N, Okamoto H, Usuki A (2004) J Appl Polym Sci 93:758
39. Usuki A, Tukigase A, Kato M (2002) Polymer 43:2185
40. Bergman JS, Chen H, Giannelis EP, Thomas MG, Coates GW (1999) Chem Commun 2179
41. Heinemann J, Reichert P, Thomann R, Mulhaupt R (1999) Macromol Rapid Comm 20:423
42. Shin S-YA, Simon LC, Soares JBP, Scholz G (2003) Polymer 44:5317
43. Wang ZM, Nakajima H, Manias E, Chung TC (2003) Macromolecules 36:8919
44. Uyama H, Kuwabara M, Tsujimoto T, Nakano M, Usuki A, Kobayashi S (2003) Chem Mater 15:2492
45. Maiti P, Yamada K, Okamoto M, Ueda K, Okamoto K (2002) Chem Mater 14:4654
46. Nam JY, Ray SS, Okamoto M (2003) Macromolecules 36:7126
47. Okamoto H, Nakano M, Ouchi M, Usuki A, Kageyama Y (2004) Mat Res Soc Symp Proc 791:399
48. Ray SS, Okamoto K, Okamoto M (2003) Macromolecules 36:2355
49. Darder M, Colilla M, Ruiz-Hitzky E (2003) Chem Mater 15:3774
50. Uyama H, Kuwabara M, Tsujimoto T, Nakano M, Usuki A, Kobayashi S (2004) Macromol Biosci 4:354
51. Yano K, Usuki A, Okada A (1997) J Polym Sci A Polym Chem 35:2289
52. Hasegawa N, Okamoto H, Kawasumi M, Usuki A (1999) J Appl Polym Sci 74:3359

Editor: Shiro Kobayashi

Author Index Volumes 101–179

Author Index Volumes 1–100 see Volume 100

de, Abajo, J. and *de la Campa, J. G.*: Processable Aromatic Polyimides. Vol. 140, pp. 23–60.
Abetz, V. see Förster, S.: Vol. 166, pp. 173–210.
Adolf, D. B. see Ediger, M. D.: Vol. 116, pp. 73–110.
Aharoni, S. M. and *Edwards, S. F.*: Rigid Polymer Networks. Vol. 118, pp. 1–231.
Albertsson, A.-C. and *Varma, I. K.*: Aliphatic Polyesters: Synthesis, Properties and Applications. Vol. 157, pp. 99–138.
Albertsson, A.-C. see Edlund, U.: Vol. 157, pp. 53–98.
Albertsson, A.-C. see Söderqvist Lindblad, M.: Vol. 157, pp. 139–161.
Albertsson, A.-C. see Stridsberg, K. M.: Vol. 157, pp. 27–51.
Albertsson, A.-C. see Al-Malaika, S.: Vol. 169, pp. 177–199.
Al-Malaika, S.: Perspectives in Stabilisation of Polyolefins. Vol. 169, pp. 121–150.
Améduri, B., Boutevin, B. and *Gramain, P.*: Synthesis of Block Copolymers by Radical Polymerization and Telomerization. Vol. 127, pp. 87–142.
Améduri, B. and *Boutevin, B.*: Synthesis and Properties of Fluorinated Telechelic Monodispersed Compounds. Vol. 102, pp. 133–170.
Amselem, S. see Domb, A. J.: Vol. 107, pp. 93–142.
Andrady, A. L.: Wavelenght Sensitivity in Polymer Photodegradation. Vol. 128, pp. 47–94.
Andreis, M. and *Koenig, J. L.*: Application of Nitrogen–15 NMR to Polymers. Vol. 124, pp. 191–238.
Angiolini, L. see Carlini, C.: Vol. 123, pp. 127–214.
Anjum, N. see Gupta, B.: Vol. 162, pp. 37–63.
Anseth, K. S., Newman, S. M. and *Bowman, C. N.*: Polymeric Dental Composites: Properties and Reaction Behavior of Multimethacrylate Dental Restorations. Vol. 122, pp. 177–218.
Antonietti, M. see Cölfen, H.: Vol. 150, pp. 67–187.
Armitage, B. A. see O'Brien, D. F.: Vol. 126, pp. 53–58.
Arndt, M. see Kaminski, W.: Vol. 127, pp. 143–187.
Arnold Jr., F. E. and *Arnold, F. E.*: Rigid-Rod Polymers and Molecular Composites. Vol. 117, pp. 257–296.
Arora, M. see Kumar, M. N. V. R.: Vol. 160, pp. 45–118.
Arshady, R.: Polymer Synthesis via Activated Esters: A New Dimension of Creativity in Macromolecular Chemistry. Vol. 111, pp. 1–42.
Auer, S. and *Frenkel, D.*: Numerical Simulation of Crystal Nucleation in Colloids. Vol. 173, pp. 149–208.

Bahar, I., Erman, B. and *Monnerie, L.*: Effect of Molecular Structure on Local Chain Dynamics: Analytical Approaches and Computational Methods. Vol. 116, pp. 145–206.
Ballauff, M. see Dingenouts, N.: Vol. 144, pp. 1–48.
Ballauff, M. see Holm, C.: Vol. 166, pp. 1–27.
Ballauff, M. see Rühe, J.: Vol. 165, pp. 79–150.

Baltá-Calleja, F. J., González Arche, A., Ezquerra, T. A., Santa Cruz, C., Batallón, F., Frick, B. and *López Cabarcos, E.*: Structure and Properties of Ferroelectric Copolymers of Poly(vinylidene) Fluoride. Vol. 108, pp. 1–48.

Baltussen, J. J. M. see Northolt, M. G.: Vol. 178, (in press)

Barnes, M. D. see Otaigbe, J. U.: Vol. 154, pp. 1–86.

Barshtein, G. R. and *Sabsai, O. Y.*: Compositions with Mineralorganic Fillers. Vol. 101, pp. 1–28.

Barton, J. see Hunkeler, D.: Vol. 112, pp. 115–134.

Baschnagel, J., Binder, K., Doruker, P., Gusev, A. A., Hahn, O., Kremer, K., Mattice, W. L., Müller-Plathe, F., Murat, M., Paul, W., Santos, S., Sutter, U. W. and *Tries, V.*:Bridging the Gap Between Atomistic and Coarse-Grained Models of Polymers: Status and Perspectives. Vol. 152, pp. 41–156.

Batallán, F. see Baltá-Calleja, F. J.: Vol. 108, pp. 1–48.

Batog, A. E., Pet'ko, I. P. and *Penczek, P.*: Aliphatic-Cycloaliphatic Epoxy Compounds and Polymers. Vol. 144, pp. 49–114.

Baughman, T. W. and *Wagener, K. B.*: Recent Advances in ADMET Polymerization. Vol 176, pp. 1–42.

Becker, O. and *Simon, G. P.*: Epoxy Layered Silicate Nanocomposites. Vol. 179, pp. 29–82.

Bell, C. L. and *Peppas, N. A.*: Biomedical Membranes from Hydrogels and Interpolymer Complexes. Vol. 122, pp. 125–176.

Bellon-Maurel, A. see Calmon-Decriaud, A.: Vol. 135, pp. 207–226.

Bennett, D. E. see O'Brien, D. F.: Vol. 126, pp. 53–84.

Berry, G. C.: Static and Dynamic Light Scattering on Moderately Concentraded Solutions: Isotropic Solutions of Flexible and Rodlike Chains and Nematic Solutions of Rodlike Chains. Vol. 114, pp. 233–290.

Bershtein, V. A. and *Ryzhov, V. A.*: Far Infrared Spectroscopy of Polymers. Vol. 114, pp. 43–122.

Bhargava, R., Wang, S.-Q. and *Koenig, J. L*: FTIR Microspectroscopy of Polymeric Systems. Vol. 163, pp. 137–191.

Biesalski, M.: see Rühe, J.: Vol. 165, pp. 79–150.

Bigg, D. M.: Thermal Conductivity of Heterophase Polymer Compositions. Vol. 119, pp. 1–30.

Binder, K.: Phase Transitions in Polymer Blends and Block Copolymer Melts: Some Recent Developments. Vol. 112, pp. 115–134.

Binder, K.: Phase Transitions of Polymer Blends and Block Copolymer Melts in Thin Films. Vol. 138, pp. 1–90.

Binder, K. see Baschnagel, J.: Vol. 152, pp. 41–156.

Binder, K., Müller, M., Virnau, P. and *González MacDowell, L.*: Polymer+Solvent Systems: Phase Diagrams, Interface Free Energies, and Nucleation. Vol. 173, pp. 1–104.

Bird, R. B. see Curtiss, C. F.: Vol. 125, pp. 1–102.

Biswas, M. and *Mukherjee, A.*: Synthesis and Evaluation of Metal-Containing Polymers. Vol. 115, pp. 89–124.

Biswas, M. and *Sinha Ray, S.*: Recent Progress in Synthesis and Evaluation of Polymer-Montmorillonite Nanocomposites. Vol. 155, pp. 167–221.

Blankenburg, L. see Klemm, E.: Vol. 177, pp. 53–90.

Bogdal, D., Penczek, P., Pielichowski, J. and *Prociak, A.*: Microwave Assisted Synthesis, Crosslinking, and Processing of Polymeric Materials. Vol. 163, pp. 193–263.

Bohrisch, J., Eisenbach, C. D., Jaeger, W., Mori, H., Müller, A. H. E., Rehahn, M., Schaller, C., Traser, S. and *Wittmeyer, P.*: New Polyelectrolyte Architectures. Vol. 165, pp. 1–41.

Bolze, J. see Dingenouts, N.: Vol. 144, pp. 1–48.

Bosshard, C.: see Gubler, U.: Vol. 158, pp. 123–190.

Boutevin, B. and *Robin, J. J.*: Synthesis and Properties of Fluorinated Diols. Vol. 102. pp. 105–132.

Boutevin, B. see Amédouri, B.: Vol. 102, pp. 133–170.

Boutevin, B. see Améduri, B.: Vol. 127, pp. 87–142.

Boutevin, B. see Guida-Pietrasanta, F.: Vol. 179, pp. 1–27.

Bowman, C. N. see Anseth, K. S.: Vol. 122, pp. 177–218.

Boyd, R. H.: Prediction of Polymer Crystal Structures and Properties. Vol. 116, pp. 1–26.

Briber, R. M. see Hedrick, J. L.: Vol. 141, pp. 1–44.

Bronnikov, S. V., Vettegren, V. I. and *Frenkel, S. Y.*: Kinetics of Deformation and Relaxation in Highly Oriented Polymers. Vol. 125, pp. 103–146.

Brown, H. R. see Creton, C.: Vol. 156, pp. 53–135.

Bruza, K. J. see Kirchhoff, R. A.: Vol. 117, pp. 1–66.

Buchmeiser, M. R.: Regioselective Polymerization of 1-Alkynes and Stereoselective Cyclopolymerization of a, w-Heptadiynes. Vol. 176, pp. 89–119.

Budkowski, A.: Interfacial Phenomena in Thin Polymer Films: Phase Coexistence and Segregation. Vol. 148, pp. 1–112.

Bunz, U. H. F.: Synthesis and Structure of PAEs. Vol. 177, pp. 1–52.

Burban, J. H. see Cussler, E. L.: Vol. 110, pp. 67–80.

Burchard, W.: Solution Properties of Branched Macromolecules. Vol. 143, pp. 113–194.

Butté, A. see Schork, F. J.: Vol. 175, pp. 129–255.

Calmon-Decriaud, A., Bellon-Maurel, V., Silvestre, F.: Standard Methods for Testing the Aerobic Biodegradation of Polymeric Materials. Vol 135, pp. 207–226.

Cameron, N. R. and *Sherrington, D. C.*: High Internal Phase Emulsions (HIPEs)-Structure, Properties and Use in Polymer Preparation. Vol. 126, pp. 163–214.

de la Campa, J. G. see de Abajo, J.: Vol. 140, pp. 23–60.

Candau, F. see Hunkeler, D.: Vol. 112, pp. 115–134.

Canelas, D. A. and *DeSimone, J. M.*: Polymerizations in Liquid and Supercritical Carbon Dioxide. Vol. 133, pp. 103–140.

Canva, M. and *Stegeman, G. I.*: Quadratic Parametric Interactions in Organic Waveguides. Vol. 158, pp. 87–121.

Capek, I.: Kinetics of the Free-Radical Emulsion Polymerization of Vinyl Chloride. Vol. 120, pp. 135–206.

Capek, I.: Radical Polymerization of Polyoxyethylene Macromonomers in Disperse Systems. Vol. 145, pp. 1–56.

Capek, I. and *Chern, C.-S.*: Radical Polymerization in Direct Mini-Emulsion Systems. Vol. 155, pp. 101–166.

Cappella, B. see Munz, M.: Vol. 164, pp. 87–210.

Carlesso, G. see Prokop, A.: Vol. 160, pp. 119–174.

Carlini, C. and *Angiolini, L.*: Polymers as Free Radical Photoinitiators. Vol. 123, pp. 127–214.

Carter, K. R. see Hedrick, J. L.: Vol. 141, pp. 1–44.

Casas-Vazquez, J. see Jou, D.: Vol. 120, pp. 207–266.

Chandrasekhar, V.: Polymer Solid Electrolytes: Synthesis and Structure. Vol 135, pp. 139–206.

Chang, J. Y. see Han, M. J.: Vol. 153, pp. 1–36.

Chang, T.: Recent Advances in Liquid Chromatography Analysis of Synthetic Polymers. Vol. 163, pp. 1–60.

Charleux, B. and *Faust, R.*: Synthesis of Branched Polymers by Cationic Polymerization. Vol. 142, pp. 1–70.

Chen, P. see Jaffe, M.: Vol. 117, pp. 297–328.

Chern, C.-S. see Capek, I.: Vol. 155, pp. 101–166.

Chevolot, Y. see Mathieu, H. J.: Vol. 162, pp. 1–35.

Choe, E.-W. see Jaffe, M.: Vol. 117, pp. 297–328.

Chow, P. Y. and *Gan, L. M.*: Microemulsion Polymerizations and Reactions. Vol. 175, pp. 257–298.

Chow, T. S.: Glassy State Relaxation and Deformation in Polymers. Vol. 103, pp. 149–190.

Chujo, Y. see Uemura, T.: Vol. 167, pp. 81–106.

Chung, S.-J. see Lin, T.-C.: Vol. 161, pp. 157–193.

Chung, T.-S. see Jaffe, M.: Vol. 117, pp. 297–328.

Cölfen, H. and *Antonietti, M.*: Field-Flow Fractionation Techniques for Polymer and Colloid Analysis. Vol. 150, pp. 67–187.

Colmenero, J. see Richter, D.: Vol. 174, in press

Comanita, B. see Roovers, J.: Vol. 142, pp. 179–228.

Connell, J. W. see Hergenrother, P. M.: Vol. 117, pp. 67–110.

Creton, C., Kramer, E. J., Brown, H. R. and *Hui, C.-Y.*: Adhesion and Fracture of Interfaces Between Immiscible Polymers: From the Molecular to the Continuum Scale. Vol. 156, pp. 53–135.

Criado-Sancho, M. see Jou, D.: Vol. 120, pp. 207–266.

Curro, J. G. see Schweizer, K. S.: Vol. 116, pp. 319–378.

Curtiss, C. F. and *Bird, R. B.*: Statistical Mechanics of Transport Phenomena: Polymeric Liquid Mixtures. Vol. 125, pp. 1–102.

Cussler, E. L., Wang, K. L. and *Burban, J. H.*: Hydrogels as Separation Agents. Vol. 110, pp. 67–80.

Dalton, L.: Nonlinear Optical Polymeric Materials: From Chromophore Design to Commercial Applications. Vol. 158, pp. 1–86.

Dautzenberg, H. see Holm, C.: Vol. 166, pp.113–171.

Davidson, J. M. see Prokop, A.: Vol. 160, pp.119–174.

Den Decker, M. G. see Northolt, M. G.: Vol. 178, (in press)

Desai, S. M. and *Singh, R. P.*: Surface Modification of Polyethylene. Vol. 169, pp. 231–293.

DeSimone, J. M. see Canelas, D. A.: Vol. 133, pp. 103–140.

DeSimone, J. M. see Kennedy, K. A.: Vol. 175, pp. 329–346.

DiMari, S. see Prokop, A.: Vol. 136, pp. 1–52.

Dimonie, M. V. see Hunkeler, D.: Vol. 112, pp. 115–134.

Dingenouts, N., Bolze, J., Pötschke, D. and *Ballauf, M.*: Analysis of Polymer Latexes by Small-Angle X-Ray Scattering. Vol. 144, pp. 1–48.

Dodd, L. R. and *Theodorou, D. N.*: Atomistic Monte Carlo Simulation and Continuum Mean Field Theory of the Structure and Equation of State Properties of Alkane and Polymer Melts. Vol. 116, pp. 249–282.

Doelker, E.: Cellulose Derivatives. Vol. 107, pp. 199–266.

Dolden, J. G.: Calculation of a Mesogenic Index with Emphasis Upon LC-Polyimides. Vol. 141, pp. 189–245.

Domb, A. J., Amselem, S., Shah, J. and *Maniar, M.*: Polyanhydrides: Synthesis and Characterization. Vol. 107, pp. 93–142.

Domb, A. J. see Kumar, M. N. V. R.: Vol. 160, pp. 45–118.

Doruker, P. see Baschnagel, J.: Vol. 152, pp. 41–156.

Dubois, P. see Mecerreyes, D.: Vol. 147, pp. 1–60.

Dubrovskii, S. A. see Kazanskii, K. S.: Vol. 104, pp. 97–134.
Dunkin, I. R. see Steinke, J.: Vol. 123, pp. 81–126.
Dunson, D. L. see McGrath, J. E.: Vol. 140, pp. 61–106.
Dziezok, P. see Rühe, J.: Vol. 165, pp. 79–150.

Eastmond, G. C.: Poly(e-caprolactone) Blends. Vol. 149, pp. 59–223.
Economy, J. and *Goranov, K.:* Thermotropic Liquid Crystalline Polymers for High Performance Applications. Vol. 117, pp. 221–256.
Ediger, M. D. and *Adolf, D. B.:* Brownian Dynamics Simulations of Local Polymer Dynamics. Vol. 116, pp. 73–110.
Edlund, U. and *Albertsson, A.-C.:* Degradable Polymer Microspheres for Controlled Drug Delivery. Vol. 157, pp. 53–98.
Edwards, S. F. see Aharoni, S. M.: Vol. 118, pp. 1–231.
Eisenbach, C. D. see Bohrisch, J.: Vol. 165, pp. 1–41.
Endo, T. see Yagci, Y.: Vol. 127, pp. 59–86.
Engelhardt, H. and *Grosche, O.:* Capillary Electrophoresis in Polymer Analysis. Vol.150, pp. 189–217.
Engelhardt, H. and *Martin, H.:* Characterization of Synthetic Polyelectrolytes by Capillary Electrophoretic Methods. Vol. 165, pp. 211–247.
Eriksson, P. see Jacobson, K.: Vol. 169, pp. 151–176.
Erman, B. see Bahar, I.: Vol. 116, pp. 145–206.
Eschner, M. see Spange, S.: Vol. 165, pp. 43–78.
Estel, K. see Spange, S.: Vol. 165, pp. 43–78.
Ewen, B. and *Richter, D.:* Neutron Spin Echo Investigations on the Segmental Dynamics of Polymers in Melts, Networks and Solutions. Vol. 134, pp. 1–130.
Ezquerra, T. A. see Baltá-Calleja, F. J.: Vol. 108, pp. 1–48.

Fatkullin, N. see Kimmich, R.: Vol. 170, pp. 1–113.
Faust, R. see Charleux, B.: Vol. 142, pp. 1–70.
Faust, R. see Kwon, Y.: Vol. 167, pp. 107–135.
Fekete, E. see Pukánszky, B.: Vol. 139, pp. 109–154.
Fendler, J. H.: Membrane-Mimetic Approach to Advanced Materials. Vol. 113, pp. 1–209.
Fetters, L. J. see Xu, Z.: Vol. 120, pp. 1–50.
Fontenot, K. see Schork, F. J.: Vol. 175, pp. 129–255.
Förster, S., Abetz, V. and *Müller, A. H. E.:* Polyelectrolyte Block Copolymer Micelles. Vol. 166, pp. 173–210.
Förster, S. and *Schmidt, M.:* Polyelectrolytes in Solution. Vol. 120, pp. 51–134.
Freire, J. J.: Conformational Properties of Branched Polymers: Theory and Simulations. Vol. 143, pp. 35–112.
Frenkel, S. Y. see Bronnikov, S. V.: Vol. 125, pp. 103–146.
Frick, B. see Baltá-Calleja, F. J.: Vol. 108, pp. 1–48.
Fridman, M. L.: see Terent'eva, J. P.: Vol. 101, pp. 29–64.
Fuchs, G. see Trimmel, G.: Vol. 176, pp. 43–87.
Fukui, K. see Otaigbe, J. U.: Vol. 154, pp. 1–86.
Funke, W.: Microgels-Intramolecularly Crosslinked Macromolecules with a Globular Structure. Vol. 136, pp. 137–232.
Furusho, Y. see Takata, T.: Vol. 171, pp. 1–75.

Galina, H.: Mean-Field Kinetic Modeling of Polymerization: The Smoluchowski Coagulation Equation. Vol. 137, pp. 135–172.

Gan, L. M. see Chow, P. Y.: Vol. 175, pp. 257–298.

Ganesh, K. see Kishore, K.: Vol. 121, pp. 81–122.

Gaw, K. O. and *Kakimoto, M.*: Polyimide-Epoxy Composites. Vol. 140, pp. 107–136.

Geckeler, K. E. see Rivas, B.: Vol. 102, pp. 171–188.

Geckeler, K. E.: Soluble Polymer Supports for Liquid-Phase Synthesis. Vol. 121, pp. 31–80.

Gedde, U. W. and *Mattozzi, A.*: Polyethylene Morphology. Vol. 169, pp. 29–73.

Gehrke, S. H.: Synthesis, Equilibrium Swelling, Kinetics Permeability and Applications of Environmentally Responsive Gels. Vol. 110, pp. 81–144.

de Gennes, P.-G.: Flexible Polymers in Nanopores. Vol. 138, pp. 91–106.

Georgiou, S.: Laser Cleaning Methodologies of Polymer Substrates. Vol. 168, pp. 1–49.

Geuss, M. see Munz, M.: Vol. 164, pp. 87–210.

Giannelis, E. P., Krishnamoorti, R. and *Manias, E.*: Polymer-Silicate Nanocomposites: Model Systems for Confined Polymers and Polymer Brushes. Vol. 138, pp. 107–148.

Godovsky, D. Y.: Device Applications of Polymer-Nanocomposites. Vol. 153, pp. 163–205.

Godovsky, D. Y.: Electron Behavior and Magnetic Properties Polymer-Nanocomposites. Vol. 119, pp. 79–122.

González Arche, A. see Baltá-Calleja, F. J.: Vol. 108, pp. 1–48.

Goranov, K. see Economy, J.: Vol. 117, pp. 221–256.

Gramain, P. see Améduri, B.: Vol. 127, pp. 87–142.

Grest, G. S.: Normal and Shear Forces Between Polymer Brushes. Vol. 138, pp. 149–184.

Grigorescu, G. and *Kulicke, W.-M.*: Prediction of Viscoelastic Properties and Shear Stability of Polymers in Solution. Vol. 152, p. 1–40.

Gröhn, F. see Rühe, J.: Vol. 165, pp. 79–150.

Grosberg, A. and *Nechaev, S.*: Polymer Topology. Vol. 106, pp. 1–30.

Grosche, O. see Engelhardt, H.: Vol. 150, pp. 189–217.

Grubbs, R., Risse, W. and *Novac, B.*: The Development of Well-defined Catalysts for Ring-Opening Olefin Metathesis. Vol. 102, pp. 47–72.

Gubler, U. and *Bosshard, C.*: Molecular Design for Third-Order Nonlinear Optics. Vol. 158, pp. 123–190.

Guida-Pietrasanta, F. and *Boutevin, B.*: Polysilalkylene or Silarylene Siloxanes Said Hybrid Silicones. Vol. 179, pp. 1–27.

van Gunsteren, W. F. see Gusev, A. A.: Vol. 116, pp. 207–248.

Gupta, B. and *Anjum, N.*: Plasma and Radiation-Induced Graft Modification of Polymers for Biomedical Applications. Vol. 162, pp. 37–63.

Gusev, A. A., Müller-Plathe, F., van Gunsteren, W. F. and *Suter, U. W.*: Dynamics of Small Molecules in Bulk Polymers. Vol. 116, pp. 207–248.

Gusev, A. A. see Baschnagel, J.: Vol. 152, pp. 41–156.

Guillot, J. see Hunkeler, D.: Vol. 112, pp. 115–134.

Guyot, A. and *Tauer, K.*: Reactive Surfactants in Emulsion Polymerization. Vol. 111, pp. 43–66.

Hadjichristidis, N., Pispas, S., Pitsikalis, M., Iatrou, H. and *Vlahos, C.*: Asymmetric Star Polymers Synthesis and Properties. Vol. 142, pp. 71–128.

Hadjichristidis, N. see Xu, Z.: Vol. 120, pp. 1–50.

Hadjichristidis, N. see Pitsikalis, M.: Vol. 135, pp. 1–138.

Hahn, O. see Baschnagel, J.: Vol. 152, pp. 41–156.

Hakkarainen, M.: Aliphatic Polyesters: Abiotic and Biotic Degradation and Degradation Products. Vol. 157, pp. 1–26.

Hakkarainen, M. and *Albertsson, A.-C.*: Environmental Degradation of Polyethylene. Vol. 169, pp. 177–199.

Hall, H. K. see Penelle, J.: Vol. 102, pp. 73–104.

Hamley, I. W.: Crystallization in Block Copolymers. Vol. 148, pp. 113–138.

Hammouda, B.: SANS from Homogeneous Polymer Mixtures: A Unified Overview. Vol. 106, pp. 87–134.

Han, M. J. and *Chang, J. Y.*: Polynucleotide Analogues. Vol. 153, pp. 1–36.

Harada, A.: Design and Construction of Supramolecular Architectures Consisting of Cyclodextrins and Polymers. Vol. 133, pp. 141–192.

Haralson, M. A. see Prokop, A.: Vol. 136, pp. 1–52.

Hasegawa, N. see Usuki, A.: Vol. 179, pp. 135–195.

Hassan, C. M. and *Peppas, N. A.*: Structure and Applications of Poly(vinyl alcohol) Hydrogels Produced by Conventional Crosslinking or by Freezing/Thawing Methods. Vol. 153, pp. 37–65.

Hawker, C. J.: Dentritic and Hyperbranched Macromolecules Precisely Controlled Macromolecular Architectures. Vol. 147, pp. 113–160.

Hawker, C. J. see Hedrick, J. L.: Vol. 141, pp. 1–44.

He, G. S. see Lin, T.-C.: Vol. 161, pp. 157–193.

Hedrick, J. L., Carter, K. R., Labadie, J. W., Miller, R. D., Volksen, W., Hawker, C. J., Yoon, D. Y., Russell, T. P., McGrath, J. E. and *Briber, R. M.*: Nanoporous Polyimides. Vol. 141, pp. 1–44.

Hedrick, J. L., Labadie, J. W., Volksen, W. and *Hilborn, J. G.*: Nanoscopically Engineered Polyimides. Vol. 147, pp. 61–112.

Hedrick, J. L. see Hergenrother, P. M.: Vol. 117, pp. 67–110.

Hedrick, J. L. see Kiefer, J.: Vol. 147, pp. 161–247.

Hedrick, J. L. see McGrath, J. E.: Vol. 140, pp. 61–106.

Heine, D. R., Grest, G. S. and *Curro, J. G.*: Structure of Polymer Melts and Blends: Comparison of Integral Equation theory and Computer Sumulation. Vol. 173, pp. 209–249.

Heinrich, G. and *Klüppel, M.*: Recent Advances in the Theory of Filler Networking in Elastomers. Vol. 160, pp. 1–44.

Heller, J.: Poly (Ortho Esters). Vol. 107, pp. 41–92.

Helm, C. A.: see Möhwald, H.: Vol. 165, pp. 151–175.

Hemielec, A. A. see Hunkeler, D.: Vol. 112, pp. 115–134.

Hergenrother, P. M., Connell, J. W., Labadie, J. W. and *Hedrick, J. L.*: Poly(arylene ether)s Containing Heterocyclic Units. Vol. 117, pp. 67–110.

Hernández-Barajas, J. see Wandrey, C.: Vol. 145, pp. 123–182.

Hervet, H. see Léger, L.: Vol. 138, pp. 185–226.

Hilborn, J. G. see Hedrick, J. L.: Vol. 147, pp. 61–112.

Hilborn, J. G. see Kiefer, J.: Vol. 147, pp. 161–247.

Hiramatsu, N. see Matsushige, M.: Vol. 125, pp. 147–186.

Hirasa, O. see Suzuki, M.: Vol. 110, pp. 241–262.

Hirotsu, S.: Coexistence of Phases and the Nature of First-Order Transition in Poly-N-isopropylacrylamide Gels. Vol. 110, pp. 1–26.

Höcker, H. see Klee, D.: Vol. 149, pp. 1–57.

Holm, C., Hofmann, T., Joanny, J. F., Kremer, K., Netz, R. R., Reineker, P., Seidel, C., Vilgis, T. A. and *Winkler, R. G.*: Polyelectrolyte Theory. Vol. 166, pp. 67–111.

Holm, C., Rehahn, M., Oppermann, W. and *Ballauff, M.*: Stiff-Chain Polyelectrolytes. Vol. 166, pp. 1–27.

Hornsby, P.: Rheology, Compounding and Processing of Filled Thermoplastics. Vol. 139, pp. 155–216.

Houbenov, N. see Rühe, J.: Vol. 165, pp. 79–150.

Huber, K. see Volk, N.: Vol. 166, pp. 29–65.

Hugenberg, N. see Rühe, J.: Vol. 165, pp. 79–150.

Hui, C.-Y. see Creton, C.: Vol. 156, pp. 53–135.

Hult, A., Johansson, M. and *Malmström, E.*: Hyperbranched Polymers. Vol. 143, pp. 1–34.

Hünenberger, P. H.: Thermostat Algorithms for Molecular-Dynamics Simulations. Vol. 173, pp. 105–147.

Hunkeler, D., Candau, F., Pichot, C., Hemielec, A. E., Xie, T. Y., Barton, J., Vaskova, V., Guillot, J., Dimonie, M. V. and *Reichert, K. H.*: Heterophase Polymerization: A Physical and Kinetic Comparision and Categorization. Vol. 112, pp. 115–134.

Hunkeler, D. see Macko, T.: Vol. 163, pp. 61–136. Hunkeler, D. see Prokop, A.: Vol. 136, pp. 1–52; 53–74. Hunkeler, D. see Wandrey, C.: Vol. 145, pp. 123–182.

Iatrou, H. see Hadjichristidis, N.: Vol. 142, pp. 71–128.

Ichikawa, T. see Yoshida, H.: Vol. 105, pp. 3–36.

Ihara, E. see Yasuda, H.: Vol. 133, pp. 53–102.

Ikada, Y. see Uyama, Y.: Vol. 137, pp. 1–40.

Ikehara, T. see Jinnuai, H.: Vol. 170, pp. 115–167.

Ilavsky, M.: Effect on Phase Transition on Swelling and Mechanical Behavior of Synthetic Hydrogels. Vol. 109, pp. 173–206.

Imai, Y.: Rapid Synthesis of Polyimides from Nylon-Salt Monomers. Vol. 140, pp. 1–23.

Inomata, H. see Saito, S.: Vol. 106, pp. 207–232.

Inoue, S. see Sugimoto, H.: Vol. 146, pp. 39–120.

Irie, M.: Stimuli-Responsive Poly(N-isopropylacrylamide), Photo- and Chemical-Induced Phase Transitions. Vol. 110, pp. 49–66.

Ise, N. see Matsuoka, H.: Vol. 114, pp. 187–232.

Ishikawa, T.: Advances in Inorganic Fibers. Vol. 178, (in press).

Ito, H.: Chemical Amplification Resists for Microlithography. Vol. 172, pp. 37–245.

Ito, K. and *Kawaguchi, S.*: Poly(macronomers), Homo- and Copolymerization. Vol. 142, pp. 129–178.

Ito, K. see Kawaguchi, S.: Vol. 175, pp. 299–328.

Ito, Y. see Suginome, M.: Vol. 171, pp. 77–136.

Ivanov, A. E. see Zubov, V. P.: Vol. 104, pp. 135–176.

Jacob, S. and *Kennedy, J.*: Synthesis, Characterization and Properties of OCTA-ARM Polyisobutylene-Based Star Polymers. Vol. 146, pp. 1–38.

Jacobson, K., Eriksson, P., Reitberger, T. and *Stenberg, B.*: Chemiluminescence as a Tool for Polyolefin. Vol. 169, pp. 151–176.

Jaeger, W. see Bohrisch, J.: Vol. 165, pp. 1–41.

Jaffe, M., Chen, P., Choe, E.-W., Chung, T.-S. and *Makhija, S.*: High Performance Polymer Blends. Vol. 117, pp. 297–328.

Jancar, J.: Structure-Property Relationships in Thermoplastic Matrices. Vol. 139, pp. 1–66.

Jen, A. K.-Y. see Kajzar, F.: Vol. 161, pp. 1–85.

Jerome, R. see Mecerreyes, D.: Vol. 147, pp. 1–60.

Jiang, M., Li, M., Xiang, M. and *Zhou, H.*: Interpolymer Complexation and Miscibility and Enhancement by Hydrogen Bonding. Vol. 146, pp. 121–194.

Jin, J. see Shim, H.-K.: Vol. 158, pp. 191–241.

Jinnai, H., Nishikawa, Y., Ikehara, T. and *Nishi, T.*: Emerging Technologies for the 3D Analysis of Polymer Structures. Vol. 170, pp. 115–167.

Jo, W. H. and *Yang, J. S.*: Molecular Simulation Approaches for Multiphase Polymer Systems. Vol. 156, pp. 1–52.

Joanny, J.-F. see Holm, C.: Vol. 166, pp. 67–111.

Joanny, J.-F. see Thünemann, A. F.: Vol. 166, pp. 113–171.

Johannsmann, D. see Rühe, J.: Vol. 165, pp. 79–150.

Johansson, M. see Hult, A.: Vol. 143, pp. 1–34.

Joos-Müller, B. see Funke, W.: Vol. 136, pp. 137–232.

Jou, D., Casas-Vazquez, J. and *Criado-Sancho, M.*: Thermodynamics of Polymer Solutions under Flow: Phase Separation and Polymer Degradation. Vol. 120, pp. 207–266.

Kaetsu, I.: Radiation Synthesis of Polymeric Materials for Biomedical and Biochemical Applications. Vol. 105, pp. 81–98.

Kaji, K. see Kanaya, T.: Vol. 154, pp. 87–141.

Kajzar, F., Lee, K.-S. and *Jen, A. K.-Y.*: Polymeric Materials and their Orientation Techniques for Second-Order Nonlinear Optics. Vol. 161, pp. 1–85.

Kakimoto, M. see Gaw, K. O.: Vol. 140, pp. 107–136.

Kaminski, W. and *Arndt, M.*: Metallocenes for Polymer Catalysis. Vol. 127, pp. 143–187.

Kammer, H. W., Kressler, H. and *Kummerloewe, C.*: Phase Behavior of Polymer Blends – Effects of Thermodynamics and Rheology. Vol. 106, pp. 31–86.

Kanaya, T. and *Kaji, K.*: Dynamcis in the Glassy State and Near the Glass Transition of Amorphous Polymers as Studied by Neutron Scattering. Vol. 154, pp. 87–141.

Kandyrin, L. B. and *Kuleznev, V. N.*: The Dependence of Viscosity on the Composition of Concentrated Dispersions and the Free Volume Concept of Disperse Systems. Vol. 103, pp. 103–148.

Kaneko, M. see Ramaraj, R.: Vol. 123, pp. 215–242.

Kang, E. T., Neoh, K. G. and *Tan, K. L.*: X-Ray Photoelectron Spectroscopic Studies of Electroactive Polymers. Vol. 106, pp. 135–190.

Karlsson, S. see Söderqvist Lindblad, M.: Vol. 157, pp. 139–161.

Karlsson, S.: Recycled Polyolefins. Material Properties and Means for Quality Determination. Vol. 169, pp. 201–229.

Kato, K. see Uyama, Y.: Vol. 137, pp. 1–40.

Kato, M. see Usuki, A.: Vol. 179, pp. 135–195.

Kautek, W. see Krüger, J.: Vol. 168, pp. 247–290.

Kawaguchi, S. see Ito, K.: Vol. 142, p 129–178.

Kawaguchi, S. and *Ito, K.*: Dispersion Polymerization. Vol. 175, pp. 299–328.

Kawata, S. see Sun, H.-B.: Vol. 170, pp. 169–273.

Kazanskii, K. S. and *Dubrovskii, S. A.*: Chemistry and Physics of Agricultural Hydrogels. Vol. 104, pp. 97–134.

Kennedy, J. P. see Jacob, S.: Vol. 146, pp. 1–38.

Kennedy, J. P. see Majoros, I.: Vol. 112, pp. 1–113.

Kennedy, K. A., Roberts, G. W. and *DeSimone, J. M.*: Heterogeneous Polymerization of Fluoroolefins in Supercritical Carbon Dioxide. Vol. 175, pp. 329–346.

Khokhlov, A., Starodybtzev, S. and *Vasilevskaya, V.*: Conformational Transitions of Polymer Gels: Theory and Experiment. Vol. 109, pp. 121–172.

Kiefer, J., Hedrick, J. L. and *Hiborn, J. G.*: Macroporous Thermosets by Chemically Induced Phase Separation. Vol. 147, pp. 161–247.

Kihara, N. see Takata, T.: Vol. 171, pp. 1–75.

Kilian, H. G. and *Pieper, T.*: Packing of Chain Segments. A Method for Describing X-Ray Patterns of Crystalline, Liquid Crystalline and Non-Crystalline Polymers. Vol. 108, pp. 49–90.

Kim, J. see Quirk, R. P.: Vol. 153, pp. 67–162.

Kim, K.-S. see Lin, T.-C.: Vol. 161, pp. 157–193.

Kimmich, R. and *Fatkullin, N.*: Polymer Chain Dynamics and NMR. Vol. 170, pp. 1–113.

Kippelen, B. and *Peyghambarian, N.*: Photorefractive Polymers and their Applications. Vol. 161, pp. 87–156.

Kirchhoff, R. A. and *Bruza, K. J.*: Polymers from Benzocyclobutenes. Vol. 117, pp. 1–66.

Kishore, K. and *Ganesh, K.*: Polymers Containing Disulfide, Tetrasulfide, Diselenide and Ditelluride Linkages in the Main Chain. Vol. 121, pp. 81–122.

Kitamaru, R.: Phase Structure of Polyethylene and Other Crystalline Polymers by Solid-State 13C/MNR. Vol. 137, pp. 41–102.

Klapper, M. see Rusanov, A. L.: Vol. 179, pp. 83–134.

Klee, D. and *Höcker, H.*: Polymers for Biomedical Applications: Improvement of the Interface Compatibility. Vol. 149, pp. 1–57.

Klemm, E., Pautzsch, T. and *Blankenburg, L.*: Organometallic PAEs. Vol. 177, pp. 53–90.

Klier, J. see Scranton, A. B.: Vol. 122, pp. 1–54.

v. Klitzing, R. and *Tieke, B.*: Polyelectrolyte Membranes. Vol. 165, pp. 177–210.

Klüppel, M.: The Role of Disorder in Filler Reinforcement of Elastomers on Various Length Scales. Vol. 164, pp. 1–86.

Klüppel, M. see Heinrich, G.: Vol. 160, pp. 1–44.

Knuuttila, H., Lehtinen, A. and *Nummila-Pakarinen, A.*: Advanced Polyethylene Technologies – Controlled Material Properties. Vol. 169, pp. 13–27.

Kobayashi, S., Shoda, S. and *Uyama, H.*: Enzymatic Polymerization and Oligomerization. Vol. 121, pp. 1–30.

Köhler, W. and *Schäfer, R.*: Polymer Analysis by Thermal-Diffusion Forced Rayleigh Scattering. Vol. 151, pp. 1–59.

Koenig, J. L. see Bhargava, R.: Vol. 163, pp. 137–191.

Koenig, J. L. see Andreis, M.: Vol. 124, pp. 191–238.

Koike, T.: Viscoelastic Behavior of Epoxy Resins Before Crosslinking. Vol. 148, pp. 139–188.

Kokko, E. see Löfgren, B.: Vol. 169, pp. 1–12.

Kokufuta, E.: Novel Applications for Stimulus-Sensitive Polymer Gels in the Preparation of Functional Immobilized Biocatalysts. Vol. 110, pp. 157–178.

Konno, M. see Saito, S.: Vol. 109, pp. 207–232.

Konradi, R. see Rühe, J.: Vol. 165, pp. 79–150.

Kopecek, J. see Putnam, D.: Vol. 122, pp. 55–124.

Koßmehl, G. see Schopf, G.: Vol. 129, pp. 1–145.

Kostoglodov, P. V. see Rusanov, A. L.: Vol. 179, pp. 83–134.

Kozlov, E. see Prokop, A.: Vol. 160, pp. 119–174.

Kramer, E. J. see Creton, C.: Vol. 156, pp. 53–135.

Kremer, K. see Baschnagel, J.: Vol. 152, pp. 41–156.

Kremer, K. see Holm, C.: Vol. 166, pp. 67–111.

Kressler, J. see Kammer, H. W.: Vol. 106, pp. 31–86.

Kricheldorf, H. R.: Liquid-Cristalline Polyimides. Vol. 141, pp. 83–188.

Krishnamoorti, R. see Giannelis, E. P.: Vol. 138, pp. 107–148.

Krüger, J. and *Kautek, W.*: Ultrashort Pulse Laser Interaction with Dielectrics and Polymers, Vol. 168, pp. 247–290.

Kuchanov, S. I.: Modern Aspects of Quantitative Theory of Free-Radical Copolymerization. Vol. 103, pp. 1–102.

Kuchanov, S. I.: Principles of Quantitive Description of Chemical Structure of Synthetic Polymers. Vol. 152, p. 157–202.

Kudaibergennow, S. E.: Recent Advances in Studying of Synthetic Polyampholytes in Solutions. Vol. 144, pp. 115–198.

Kuleznev, V. N. see Kandyrin, L. B.: Vol. 103, pp. 103–148.

Kulichkhin, S. G. see Malkin, A. Y.: Vol. 101, pp. 217–258.

Kulicke, W.-M. see Grigorescu, G.: Vol. 152, p. 1–40.

Kumar, M. N. V. R., Kumar, N., Domb, A. J. and *Arora, M.:* Pharmaceutical Polymeric Controlled Drug Delivery Systems. Vol. 160, pp. 45–118.

Kumar, N. see Kumar, M. N. V. R.: Vol. 160, pp. 45–118.

Kummerloewe, C. see Kammer, H. W.: Vol. 106, pp. 31–86.

Kuznetsova, N. P. see Samsonov, G. V.: Vol. 104, pp. 1–50.

Kwon, Y. and *Faust, R.:* Synthesis of Polyisobutylene-Based Block Copolymers with Precisely Controlled Architecture by Living Cationic Polymerization. Vol. 167, pp. 107–135.

Labadie, J. W. see Hergenrother, P. M.: Vol. 117, pp. 67–110.

Labadie, J. W. see Hedrick, J. L.: Vol. 141, pp. 1–44.

Labadie, J. W. see Hedrick, J. L.: Vol. 147, pp. 61–112.

Lamparski, H. G. see O'Brien, D. F.: Vol. 126, pp. 53–84.

Laschewsky, A.: Molecular Concepts, Self-Organisation and Properties of Polysoaps. Vol. 124, pp. 1–86.

Laso, M. see Leontidis, E.: Vol. 116, pp. 283–318.

Lazár, M. and *Rychlý, R.:* Oxidation of Hydrocarbon Polymers. Vol. 102, pp. 189–222.

Lechowicz, J. see Galina, H.: Vol. 137, pp. 135–172.

Léger, L., Raphaël, E. and *Hervet, H.:* Surface-Anchored Polymer Chains: Their Role in Adhesion and Friction. Vol. 138, pp. 185–226.

Lenz, R. W.: Biodegradable Polymers. Vol. 107, pp. 1–40.

Leontidis, E., de Pablo, J. J., Laso, M. and *Suter, U. W.:* A Critical Evaluation of Novel Algorithms for the Off-Lattice Monte Carlo Simulation of Condensed Polymer Phases. Vol. 116, pp. 283–318.

Lee, B. see Quirk, R. P.: Vol. 153, pp. 67–162.

Lee, K.-S. see Kajzar, F.: Vol. 161, pp. 1–85.

Lee, Y. see Quirk, R. P: Vol. 153, pp. 67–162.

Lehtinen, A. see Knuuttila, H.: Vol. 169, pp. 13–27.

Leónard, D. see Mathieu, H. J.: Vol. 162, pp. 1–35.

Lesec, J. see Viovy, J.-L.: Vol. 114, pp. 1–42.

Li, M. see Jiang, M.: Vol. 146, pp. 121–194.

Liang, G. L. see Sumpter, B. G.: Vol. 116, pp. 27–72.

Lienert, K.-W.: Poly(ester-imide)s for Industrial Use. Vol. 141, pp. 45–82.

Likhatchev, D. see Rusanov, A. L.: Vol. 179, pp. 83–134.

Lin, J. and *Sherrington, D. C.:* Recent Developments in the Synthesis, Thermostability and Liquid Crystal Properties of Aromatic Polyamides. Vol. 111, pp. 177–220.

Lin, T.-C., Chung, S.-J., Kim, K.-S., Wang, X., He, G. S., Swiatkiewicz, J., Pudavar, H. E. and *Prasad, P. N.:* Organics and Polymers with High Two-Photon Activities and their Applications. Vol. 161, pp. 157–193.

Lippert, T.: Laser Application of Polymers. Vol. 168, pp. 51–246.

Liu, Y. see Söderqvist Lindblad, M.: Vol. 157, pp. 139–161.

López Cabarcos, E. see Baltá-Calleja, F. J.: Vol. 108, pp. 1–48.

Löfgren, B., Kokko, E. and *Seppälä, J.:* Specific Structures Enabled by Metallocene Catalysis in Polyethenes. Vol. 169, pp. 1–12.

Löwen, H. see Thünemann, A. F.: Vol. 166, pp. 113–171.

Luo, Y. see Schork, F. J.: Vol. 175, pp. 129–255.

Macko, T. and *Hunkeler, D.:* Liquid Chromatography under Critical and Limiting Conditions: A Survey of Experimental Systems for Synthetic Polymers. Vol. 163, pp. 61–136.

Majoros, I., Nagy, A. and *Kennedy, J. P.*: Conventional and Living Carbocationic Polymerizations United. I. A Comprehensive Model and New Diagnostic Method to Probe the Mechanism of Homopolymerizations. Vol. 112, pp. 1–113.

Makhija, S. see Jaffe, M.: Vol. 117, pp. 297–328.

Malmström, E. see Hult, A.: Vol. 143, pp. 1–34.

Malkin, A. Y. and *Kulichkhin, S. G.*: Rheokinetics of Curing. Vol. 101, pp. 217–258.

Maniar, M. see Domb, A. J.: Vol. 107, pp. 93–142.

Manias, E. see Giannelis, E. P.: Vol. 138, pp. 107–148.

Martin, H. see Engelhardt, H.: Vol. 165, pp. 211–247.

Marty, J. D. and *Mauzac, M.*: Molecular Imprinting: State of the Art and Perspectives. Vol. 172, pp. 1–35.

Mashima, K., Nakayama, Y. and *Nakamura, A.*: Recent Trends in Polymerization of a-Olefins Catalyzed by Organometallic Complexes of Early Transition Metals. Vol. 133, pp. 1–52.

Mathew, D. see Reghunadhan Nair, C. P.: Vol. 155, pp. 1–99.

Mathieu, H. J., Chevolot, Y, Ruiz-Taylor, L. and *Leónard, D.*: Engineering and Characterization of Polymer Surfaces for Biomedical Applications. Vol. 162, pp. 1–35.

Matsumoto, A.: Free-Radical Crosslinking Polymerization and Copolymerization of Multivinyl Compounds. Vol. 123, pp. 41–80.

Matsumoto, A. see Otsu, T.: Vol. 136, pp. 75–138.

Matsuoka, H. and *Ise, N.*: Small-Angle and Ultra-Small Angle Scattering Study of the Ordered Structure in Polyelectrolyte Solutions and Colloidal Dispersions. Vol. 114, pp. 187–232.

Matsushige, K., Hiramatsu, N. and *Okabe, H.*: Ultrasonic Spectroscopy for Polymeric Materials. Vol. 125, pp. 147–186.

Mattice, W. L. see Rehahn, M.: Vol. 131/132, pp. 1–475.

Mattice, W. L. see Baschnagel, J.: Vol. 152, pp. 41–156.

Mattozzi, A. see Gedde, U. W.: Vol. 169, pp. 29–73.

Mauzac, M. see Marty, J. D.: Vol. 172, pp. 1–35.

Mays, W. see Xu, Z.: Vol. 120, pp. 1–50.

Mays, J. W. see Pitsikalis, M.: Vol. 135, pp. 1–138.

McGrath, J. E. see Hedrick, J. L.: Vol. 141, pp. 1–44.

McGrath, J. E., Dunson, D. L. and *Hedrick, J. L.*: Synthesis and Characterization of Segmented Polyimide-Polyorganosiloxane Copolymers. Vol. 140, pp. 61–106.

McLeish, T. C. B. and *Milner, S. T.*: Entangled Dynamics and Melt Flow of Branched Polymers. Vol. 143, pp. 195–256.

Mecerreyes, D., Dubois, P. and *Jerome, R.*: Novel Macromolecular Architectures Based on Aliphatic Polyesters: Relevance of the Coordination-Insertion Ring-Opening Polymerization. Vol. 147, pp. 1–60.

Mecham, S. J. see McGrath, J. E.: Vol. 140, pp. 61–106.

Menzel, H. see Möhwald, H.: Vol. 165, pp. 151–175.

Meyer, T. see Spange, S.: Vol. 165, pp. 43–78.

Mikos, A. G. see Thomson, R. C.: Vol. 122, pp. 245–274.

Milner, S. T. see McLeish, T. C. B.: Vol. 143, pp. 195–256.

Mison, P. and *Sillion, B.*: Thermosetting Oligomers Containing Maleimides and Nadiimides End-Groups. Vol. 140, pp. 137–180.

Miyasaka, K.: PVA-Iodine Complexes: Formation, Structure and Properties. Vol. 108. pp. 91–130.

Miller, R. D. see Hedrick, J. L.: Vol. 141, pp. 1–44.

Minko, S. see Rühe, J.: Vol. 165, pp. 79–150.

Möhwald, H., Menzel, H., Helm, C. A. and *Stamm, M.*: Lipid and Polyampholyte Monolayers to Study Polyelectrolyte Interactions and Structure at Interfaces. Vol. 165, pp. 151–175.
Monkenbusch, M. see Richter, D.: Vol. 174, in press
Monnerie, L. see Bahar, I.: Vol. 116, pp. 145–206.
Moore, J. S. see Ray, C. R.: Vol. 177, pp. 99–149.
Mori, H. see Bohrisch, J.: Vol. 165, pp. 1–41.
Morishima, Y.: Photoinduced Electron Transfer in Amphiphilic Polyelectrolyte Systems. Vol. 104, pp. 51–96.
Morton, M. see Quirk, R. P: Vol. 153, pp. 67–162.
Motornov, M. see Rühe, J.: Vol. 165, pp. 79–150.
Mours, M. see Winter, H. H.: Vol. 134, pp. 165–234.
Müllen, K. see Scherf, U.: Vol. 123, pp. 1–40.
Müller, A. H. E. see Bohrisch, J.: Vol. 165, pp. 1–41.
Müller, A. H. E. see Förster, S.: Vol. 166, pp. 173–210.
Müller, M. see Thünemann, A. F.: Vol. 166, pp. 113–171.
Müller-Plathe, F. see Gusev, A. A.: Vol. 116, pp. 207–248.
Müller-Plathe, F. see Baschnagel, J.: Vol. 152, p. 41–156.
Mukerherjee, A. see Biswas, M.: Vol. 115, pp. 89–124.
Munz, M., Cappella, B., Sturm, H., Geuss, M. and *Schulz, E.*: Materials Contrasts and Nanolithography Techniques in Scanning Force Microscopy (SFM) and their Application to Polymers and Polymer Composites. Vol. 164, pp. 87–210.
Murat, M. see Baschnagel, J.: Vol. 152, p. 41–156.
Mylnikov, V.: Photoconducting Polymers. Vol. 115, pp. 1–88.

Nagy, A. see Majoros, I.: Vol. 112, pp. 1–11.
Naka, K. see Uemura, T.: Vol. 167, pp. 81–106.
Nakamura, A. see Mashima, K.: Vol. 133, pp. 1–52.
Nakayama, Y. see Mashima, K.: Vol. 133, pp. 1–52.
Narasinham, B. and *Peppas, N. A.*: The Physics of Polymer Dissolution: Modeling Approaches and Experimental Behavior. Vol. 128, pp. 157–208.
Nechaev, S. see Grosberg, A.: Vol. 106, pp. 1–30.
Neoh, K. G. see Kang, E. T.: Vol. 106, pp. 135–190.
Netz, R. R. see Holm, C.: Vol. 166, pp. 67–111.
Netz, R. R. see Rühe, J.: Vol. 165, pp. 79–150.
Newman, S. M. see Anseth, K. S.: Vol. 122, pp. 177–218.
Nijenhuis, K. te: Thermoreversible Networks. Vol. 130, pp. 1–252.
Ninan, K. N. see Reghunadhan Nair, C. P.: Vol. 155, pp. 1–99.
Nishi, T. see Jinnai, H.: Vol. 170, pp. 115–167.
Nishikawa, Y. see Jinnai, H.: Vol. 170, pp. 115–167.
Noid, D. W. see Otaigbe, J. U.: Vol. 154, pp. 1–86.
Noid, D. W. see Sumpter, B. G.: Vol. 116, pp. 27–72.
Nomura, M., Tobita, H. and *Suzuki, K.*: Emulsion Polymerization: Kinetic and Mechanistic Aspects. Vol. 175, pp. 1–128.
Northolt, M. G., Picken, S. J., Den Decker, M. G., Baltussen, J. J. M. and *Schlatmann, R.*: The Tensile Strength of Polymer Fibres. Vol 178, (in press).
Novac, B. see Grubbs, R.: Vol. 102, pp. 47–72.
Novikov, V. V. see Privalko, V. P.: Vol. 119, pp. 31–78.
Nummila-Pakarinen, A. see Knuuttila, H.: Vol. 169, pp. 13–27.

O'Brien, D. F., Armitage, B. A., Bennett, D. E. and *Lamparski, H. G.*: Polymerization and Domain Formation in Lipid Assemblies. Vol. 126, pp. 53–84.

Ogasawara, M.: Application of Pulse Radiolysis to the Study of Polymers and Polymerizations. Vol.105, pp. 37–80.

Okabe, H. see Matsushige, K.: Vol. 125, pp. 147–186.

Okada, M.: Ring-Opening Polymerization of Bicyclic and Spiro Compounds. Reactivities and Polymerization Mechanisms. Vol. 102, pp. 1–46.

Okano, T.: Molecular Design of Temperature-Responsive Polymers as Intelligent Materials. Vol. 110, pp. 179–198.

Okay, O. see Funke, W.: Vol. 136, pp. 137–232.

Onuki, A.: Theory of Phase Transition in Polymer Gels. Vol. 109, pp. 63–120.

Oppermann, W. see Holm, C.: Vol. 166, pp. 1–27.

Oppermann, W. see Volk, N.: Vol. 166, pp. 29–65.

Osad'ko, I. S.: Selective Spectroscopy of Chromophore Doped Polymers and Glasses. Vol. 114, pp. 123–186.

Osakada, K. and *Takeuchi, D.*: Coordination Polymerization of Dienes, Allenes, and Methylenecycloalkanes. Vol. 171, pp. 137–194.

Otaigbe, J. U., Barnes, M. D., Fukui, K., Sumpter, B. G. and *Noid, D. W.*: Generation, Characterization, and Modeling of Polymer Micro- and Nano-Particles. Vol. 154, pp. 1–86.

Otsu, T. and *Matsumoto, A.*: Controlled Synthesis of Polymers Using the Iniferter Technique: Developments in Living Radical Polymerization. Vol. 136, pp. 75–138.

de Pablo, J. J. see Leontidis, E.: Vol. 116, pp. 283–318.

Padias, A. B. see Penelle, J.: Vol. 102, pp. 73–104.

Pascault, J.-P. see Williams, R. J. J.: Vol. 128, pp. 95–156.

Pasch, H.: Analysis of Complex Polymers by Interaction Chromatography. Vol. 128, pp. 1–46.

Pasch, H.: Hyphenated Techniques in Liquid Chromatography of Polymers. Vol. 150, pp. 1–66.

Paul, W. see Baschnagel, J.: Vol. 152, p. 41–156.

Pautzsch, T. see Klemm, E.: Vol. 177, pp. 53–90.

Penczek, P. see Batog, A. E.: Vol. 144, pp. 49–114.

Penczek, P. see Bogdal, D.: Vol. 163, pp. 193–263.

Penelle, J., Hall, H. K., Padias, A. B. and *Tanaka, H.*: Captodative Olefins in Polymer Chemistry. Vol. 102, pp. 73–104.

Peppas, N. A. see Bell, C. L.: Vol. 122, pp. 125–176.

Peppas, N. A. see Hassan, C. M.: Vol. 153, pp. 37–65.

Peppas, N. A. see Narasimhan, B.: Vol. 128, pp. 157–208.

Pet'ko, I. P. see Batog, A. E.: Vol. 144, pp. 49–114.

Pheyghambarian, N. see Kippelen, B.: Vol. 161, pp. 87–156.

Pichot, C. see Hunkeler, D.: Vol. 112, pp. 115–134.

Picken, S. J. see Northolt, M. G.: Vol. 178, (in press)

Pielichowski, J. see Bogdal, D.: Vol. 163, pp. 193–263.

Pieper, T. see Kilian, H. G.: Vol. 108, pp. 49–90.

Pispas, S. see Pitsikalis, M.: Vol. 135, pp. 1–138.

Pispas, S. see Hadjichristidis, N.: Vol. 142, pp. 71–128.

Pitsikalis, M., Pispas, S., Mays, J. W. and *Hadjichristidis, N.*: Nonlinear Block Copolymer Architectures. Vol. 135, pp. 1–138.

Pitsikalis, M. see Hadjichristidis, N.: Vol. 142, pp. 71–128.

Pleul, D. see Spange, S.: Vol. 165, pp. 43–78.

Plummer, C. J. G.: Microdeformation and Fracture in Bulk Polyolefins. Vol. 169, pp. 75–119.

Pötschke, D. see Dingenouts, N.: Vol 144, pp. 1–48.

Pokrovskii, V. N.: The Mesoscopic Theory of the Slow Relaxation of Linear Macromolecules. Vol. 154, pp. 143–219.

Pospíšil, J.: Functionalized Oligomers and Polymers as Stabilizers for Conventional Polymers. Vol. 101, pp. 65–168.

Pospíšil, J.: Aromatic and Heterocyclic Amines in Polymer Stabilization. Vol. 124, pp. 87–190.

Powers, A. C. see Prokop, A.: Vol. 136, pp. 53–74.

Prasad, P. N. see Lin, T.-C.: Vol. 161, pp. 157–193.

Priddy, D. B.: Recent Advances in Styrene Polymerization. Vol. 111, pp. 67–114.

Priddy, D. B.: Thermal Discoloration Chemistry of Styrene-co-Acrylonitrile. Vol. 121, pp. 123–154.

Privalko, V. P. and *Novikov, V. V.*: Model Treatments of the Heat Conductivity of Heterogeneous Polymers. Vol. 119, pp. 31–78.

Prociak, A. see Bogdal, D.: Vol. 163, pp. 193–263.

Prokop, A., Hunkeler, D., DiMari, S., Haralson, M. A. and *Wang, T. G.*: Water Soluble Polymers for Immunoisolation I: Complex Coacervation and Cytotoxicity. Vol. 136, pp. 1–52.

Prokop, A., Hunkeler, D., Powers, A. C., Whitesell, R. R. and *Wang, T. G.*: Water Soluble Polymers for Immunoisolation II: Evaluation of Multicomponent Microencapsulation Systems. Vol. 136, pp. 53–74.

Prokop, A., Kozlov, E., Carlesso, G. and *Davidsen, J. M.*: Hydrogel-Based Colloidal Polymeric System for Protein and Drug Delivery: Physical and Chemical Characterization, Permeability Control and Applications. Vol. 160, pp. 119–174.

Pruitt, L. A.: The Effects of Radiation on the Structural and Mechanical Properties of Medical Polymers. Vol. 162, pp. 65–95.

Pudavar, H. E. see Lin, T.-C.: Vol. 161, pp. 157–193.

Pukánszky, B. and *Fekete, E.*: Adhesion and Surface Modification. Vol. 139, pp. 109–154.

Putnam, D. and *Kopecek, J.*: Polymer Conjugates with Anticancer Acitivity. Vol. 122, pp. 55–124.

Quirk, R. P., Yoo, T., Lee, Y., M., Kim, J. and *Lee, B.*: Applications of 1,1-Diphenylethylene Chemistry in Anionic Synthesis of Polymers with Controlled Structures. Vol. 153, pp. 67–162.

Ramaraj, R. and *Kaneko, M.*: Metal Complex in Polymer Membrane as a Model for Photosynthetic Oxygen Evolving Center. Vol. 123, pp. 215–242.

Rangarajan, B. see Scranton, A. B.: Vol. 122, pp. 1–54.

Ranucci, E. see Söderqvist Lindblad, M.: Vol. 157, pp. 139–161.

Raphaël, E. see Léger, L.: Vol. 138, pp. 185–226.

Ray, C. R. and *Moore, J. S.*: Supramolecular Organization of Foldable Phenylene Ethynylene Oligomers. Vol. 177, pp. 99–149.

Reddinger, J. L. and *Reynolds, J. R.*: Molecular Engineering of p-Conjugated Polymers. Vol. 145, pp. 57–122.

Reghunadhan Nair, C. P., Mathew, D. and *Ninan, K. N.*: Cyanate Ester Resins, Recent Developments. Vol. 155, pp. 1–99.

Reichert, K. H. see Hunkeler, D.: Vol. 112, pp. 115–134.

Rehahn, M., Mattice, W. L. and *Suter, U. W.*: Rotational Isomeric State Models in Macromolecular Systems. Vol. 131/132, pp. 1–475.

Rehahn, M. see Bohrisch, J.: Vol. 165, pp. 1–41.

Rehahn, M. see Holm, C.: Vol. 166, pp. 1–27.

Reineker, P. see Holm, C.: Vol. 166, pp. 67–111.

Reitberger, T. see Jacobson, K.: Vol. 169, pp. 151–176.

Reynolds, J. R. see Reddinger, J. L.: Vol. 145, pp. 57–122.

Richter, D. see Ewen, B.: Vol. 134, pp.1–130.

Richter, D., Monkenbusch, M. and *Colmenero, J.*: Neutron Spin Echo in Polymer Systems. Vol. 174, in press

Riegler, S. see Trimmel, G.: Vol. 176, pp. 43–87.

Risse, W. see Grubbs, R.: Vol. 102, pp. 47–72.

Rivas, B. L. and *Geckeler, K. E.*: Synthesis and Metal Complexation of Poly(ethyleneimine) and Derivatives. Vol. 102, pp. 171–188.

Roberts, G. W. see Kennedy, K. A.: Vol. 175, pp. 329–346.

Robin, J. J.: The Use of Ozone in the Synthesis of New Polymers and the Modification of Polymers. Vol. 167, pp. 35–79.

Robin, J. J. see Boutevin, B.: Vol. 102, pp. 105–132.

Roe, R.-J.: MD Simulation Study of Glass Transition and Short Time Dynamics in Polymer Liquids. Vol. 116, pp. 111–114.

Roovers, J. and *Comanita, B.*: Dendrimers and Dendrimer-Polymer Hybrids. Vol. 142, pp. 179–228.

Rothon, R. N.: Mineral Fillers in Thermoplastics: Filler Manufacture and Characterisation. Vol. 139, pp. 67–108.

Rozenberg, B. A. see Williams, R. J. J.: Vol. 128, pp. 95–156.

Rühe, J., Ballauff, M., Biesalski, M., Dziezok, P., Gröhn, F., Johannsmann, D., Houbenov, N., Hugenberg, N., Konradi, R., Minko, S., Motornov, M., Netz, R. R., Schmidt, M., Seidel, C., Stamm, M., Stephan, T., Usov, D. and *Zhang, H.*: Polyelectrolyte Brushes. Vol. 165, pp. 79–150.

Ruckenstein, E.: Concentrated Emulsion Polymerization. Vol. 127, pp. 1–58.

Ruiz-Taylor, L. see Mathieu, H. J.: Vol. 162, pp. 1–35.

Rusanov, A. L.: Novel Bis (Naphtalic Anhydrides) and Their Polyheteroarylenes with Improved Processability. Vol. 111, pp. 115–176.

Rusanov, A. L., Likhatchev, D., Kostoglodov, P. V., Müllen, K. and *Klapper, M.*: Proton-Exchanging Electrolyte Membranes Based on Aromatic Condensation Polymers. Vol. 179, pp. 83–134.

Russel, T. P. see Hedrick, J. L.: Vol. 141, pp. 1–44.

Russum, J. P. see Schork, F. J.: Vol. 175, pp. 129–255.

Rychly, J. see Lazár, M.: Vol. 102, pp. 189–222.

Ryner, M. see Stridsberg, K. M.: Vol. 157, pp. 27–51.

Ryzhov, V. A. see Bershtein, V. A.: Vol. 114, pp. 43–122.

Sabsai, O. Y. see Barshtein, G. R.: Vol. 101, pp. 1–28.

Saburov, V. V. see Zubov, V. P.: Vol. 104, pp. 135–176.

Saito, S., Konno, M. and *Inomata, H.*: Volume Phase Transition of N-Alkylacrylamide Gels. Vol. 109, pp. 207–232.

Samsonov, G. V. and *Kuznetsova, N. P.*: Crosslinked Polyelectrolytes in Biology. Vol. 104, pp. 1–50.

Santa Cruz, C. see Baltá-Calleja, F. J.: Vol. 108, pp. 1–48.

Santos, S. see Baschnagel, J.: Vol. 152, p. 41–156.

Sato, T. and *Teramoto, A.*: Concentrated Solutions of Liquid-Christalline Polymers. Vol. 126, pp. 85–162.

Schaller, C. see Bohrisch, J.: Vol. 165, pp. 1–41.

Schäfer, R. see Köhler, W.: Vol. 151, pp. 1–59.

Scherf, U. and *Müllen, K.*: The Synthesis of Ladder Polymers. Vol. 123, pp. 1–40.

Schlatmann, R. see Northolt, M. G.: Vol. 178, (in press)

Schmidt, M. see Förster, S.: Vol. 120, pp. 51–134.

Schmidt, M. see Rühe, J.: Vol. 165, pp. 79–150.

Schmidt, M. see Volk, N.: Vol. 166, pp. 29–65.

Scholz, M.: Effects of Ion Radiation on Cells and Tissues. Vol. 162, pp. 97–158.

Schopf, G. and *Koßmehl, G.*: Polythiophenes – Electrically Conductive Polymers. Vol. 129, pp. 1–145.

Schork, F. J., Luo, Y., Smulders, W., Russum, J. P., Butté, A. and *Fontenot, K.*: Miniemulsion Polymerization. Vol. 175, pp. 127–255.

Schulz, E. see Munz, M.: Vol. 164, pp. 97–210.

Seppälä, J. see Löfgren, B.: Vol. 169, pp. 1–12.

Sturm, H. see Munz, M.: Vol. 164, pp. 87–210.

Schweizer, K. S.: Prism Theory of the Structure, Thermodynamics, and Phase Transitions of Polymer Liquids and Alloys. Vol. 116, pp. 319–378.

Scranton, A. B., Rangarajan, B. and *Klier, J.*: Biomedical Applications of Polyelectrolytes. Vol. 122, pp. 1–54.

Sefton, M. V. and *Stevenson, W. T. K.*: Microencapsulation of Live Animal Cells Using Polycrylates. Vol. 107, pp. 143–198.

Seidel, C. see Holm, C.: Vol. 166, pp. 67–111.

Seidel, C. see Rühe, J.: Vol. 165, pp. 79–150.

Shamanin, V. V.: Bases of the Axiomatic Theory of Addition Polymerization. Vol. 112, pp. 135–180.

Sheiko, S. S.: Imaging of Polymers Using Scanning Force Microscopy: From Superstructures to Individual Molecules. Vol. 151, pp. 61–174.

Sherrington, D. C. see Cameron, N. R.: Vol. 126, pp. 163–214.

Sherrington, D. C. see Lin, J.: Vol. 111, pp. 177–220.

Sherrington, D. C. see Steinke, J.: Vol. 123, pp. 81–126.

Shibayama, M. see Tanaka, T.: Vol. 109, pp. 1–62.

Shiga, T.: Deformation and Viscoelastic Behavior of Polymer Gels in Electric Fields. Vol. 134, pp. 131–164.

Shim, H.-K. and *Jin, J.*: Light-Emitting Characteristics of Conjugated Polymers. Vol. 158, pp. 191–241.

Shoda, S. see Kobayashi, S.: Vol. 121, pp. 1–30.

Siegel, R. A.: Hydrophobic Weak Polyelectrolyte Gels: Studies of Swelling Equilibria and Kinetics. Vol. 109, pp. 233–268.

Silvestre, F. see Calmon-Decriaud, A.: Vol. 207, pp. 207–226.

Sillion, B. see Mison, P.: Vol. 140, pp. 137–180.

Simon, F. see Spange, S.: Vol. 165, pp. 43–78.

Simon, G. P. see Becker, O.: Vol. 179, pp. 29–82.

Singh, R. P. see Sivaram, S.: Vol. 101, pp. 169–216.

Singh, R. P. see Desai, S. M.: Vol. 169, pp. 231–293.

Sinha Ray, S. see Biswas, M: Vol. 155, pp. 167–221.

Sivaram, S. and *Singh, R. P.*: Degradation and Stabilization of Ethylene-Propylene Copolymers and Their Blends: A Critical Review. Vol. 101, pp. 169–216.

Slugovc, C. see Trimmel, G.: Vol. 176, pp. 43–87.

Smulders, W. see Schork, F. J.: Vol. 175, pp. 129–255.

Söderqvist Lindblad, M., Liu, Y., Albertsson, A.-C., Ranucci, E. and *Karlsson, S.*:Polymer from Renewable Resources. Vol. 157, pp. 139–161.

Spange, S., Meyer, T., Voigt, I., Eschner, M., Estel, K., Pleul, D. and *Simon, F.*:Poly(Vinyl-formamide-co-Vinylamine)/Inorganic Oxid Hybrid Materials. Vol. 165, pp. 43–78.

Stamm, M. see Möhwald, H.: Vol. 165, pp. 151–175.

Stamm, M. see Rühe, J.: Vol. 165, pp. 79–150.

Starodybtzev, S. see Khokhlov, A.: Vol. 109, pp. 121–172.

Stegeman, G. I. see Canva, M.: Vol. 158, pp. 87–121.

Steinke, J., Sherrington, D. C. and *Dunkin, I. R.*: Imprinting of Synthetic Polymers Using Molecular Templates. Vol. 123, pp. 81–126.

Stelzer, F. see Trimmel, G.: Vol. 176, pp. 43–87.

Stenberg, B. see Jacobson, K.: Vol. 169, pp. 151–176.

Stenzenberger, H. D.: Addition Polyimides. Vol. 117, pp. 165–220.

Stephan, T. see Rühe, J.: Vol. 165, pp. 79–150.

Stevenson, W. T. K. see Sefton, M. V.: Vol. 107, pp. 143–198.

Stridsberg, K. M., Ryner, M. and *Albertsson, A.-C.*: Controlled Ring-Opening Polymerization: Polymers with Designed Macromoleculars Architecture. Vol. 157, pp. 27–51.

Sturm, H. see Munz, M.: Vol. 164, pp. 87–210.

Suematsu, K.: Recent Progress of Gel Theory: Ring, Excluded Volume, and Dimension. Vol. 156, pp. 136–214.

Sugimoto, H. and *Inoue, S.*: Polymerization by Metalloporphyrin and Related Complexes. Vol. 146, pp. 39–120.

Suginome, M. and *Ito, Y.*: Transition Metal-Mediated Polymerization of Isocyanides. Vol. 171, pp. 77–136.

Sumpter, B. G., Noid, D. W., Liang, G. L. and *Wunderlich, B.*: Atomistic Dynamics of Macro-molecular Crystals. Vol. 116, pp. 27–72.

Sumpter, B. G. see Otaigbe, J. U.: Vol. 154, pp. 1–86.

Sun, H.-B. and *Kawata, S.*: Two-Photon Photopolymerization and 3D Lithographic Micro-fabrication. Vol. 170, pp. 169–273.

Suter, U. W. see Gusev, A. A.: Vol. 116, pp. 207–248.

Suter, U. W. see Leontidis, E.: Vol. 116, pp. 283–318.

Suter, U. W. see Rehahn, M.: Vol. 131/132, pp. 1–475.

Suter, U. W. see Baschnagel, J.: Vol. 152, p. 41–156.

Suzuki, A.: Phase Transition in Gels of Sub-Millimeter Size Induced by Interaction with Stimuli. Vol. 110, pp. 199–240. Suzuki, A. and Hirasa, O.: An Approach to Artifical Muscle by Polymer Gels due to Micro-

Phase Separation. Vol. 110, pp. 241–262. Suzuki, K. see Nomura, M.: Vol. 175, pp. 1–128. Swiatkiewicz, J. see Lin, T.-C.: Vol. 161, pp. 157–193.

Tagawa, S.: Radiation Effects on Ion Beams on Polymers. Vol. 105, pp. 99–116.

Takata, T., Kihara, N. and *Furusho, Y.*: Polyrotaxanes and Polycatenanes: Recent Advances in Syntheses and Applications of Polymers Comprising of Interlocked Structures. Vol. 171, pp. 1–75.

Takeuchi, D. see Osakada, K.: Vol. 171, pp. 137–194.

Tan, K. L. see Kang, E. T.: Vol. 106, pp. 135–190.

Tanaka, H. and *Shibayama, M.*: Phase Transition and Related Phenomena of Polymer Gels. Vol. 109, pp. 1–62.

Tanaka, T. see Penelle, J.: Vol. 102, pp. 73–104.

Tauer, K. see Guyot, A.: Vol. 111, pp. 43–66.

Teramoto, A. see Sato, T.: Vol. 126, pp. 85–162.

Terent'eva, J. P. and *Fridman, M. L.*: Compositions Based on Aminoresins. Vol. 101, pp. 29–64.

Theodorou, D. N. see Dodd, L. R.: Vol. 116, pp. 249–282.

Thomson, R. C., Wake, M. C., Yaszemski, M. J. and *Mikos, A. G.*: Biodegradable Polymer Scaffolds to Regenerate Organs. Vol. 122, pp. 245–274.

Thünemann, A. F., Müller, M., Dautzenberg, H., Joanny, J.-F. and *Löwen, H.*: Polyelectrolyte complexes. Vol. 166, pp. 113–171.

Tieke, B. see v. Klitzing, R.: Vol. 165, pp. 177–210.

Tobita, H. see Nomura, M.: Vol. 175, pp. 1–128.

Tokita, M.: Friction Between Polymer Networks of Gels and Solvent. Vol. 110, pp. 27–48.

Traser, S. see Bohrisch, J.: Vol. 165, pp. 1–41.

Tries, V. see Baschnagel, J.: Vol. 152, p. 41–156.

Trimmel, G., Riegler, S., Fuchs, G., Slugovc, C. and *Stelzer, F.*: Liquid Crystalline Polymers by Metathesis Polymerization. Vol. 176, pp. 43–87.

Tsuruta, T.: Contemporary Topics in Polymeric Materials for Biomedical Applications. Vol. 126, pp. 1–52.

Uemura, T., Naka, K. and *Chujo, Y.*: Functional Macromolecules with Electron-Donating
Dithiafulvene Unit. Vol. 167, pp. 81–106. Usov, D. see Rühe, J.: Vol. 165, pp. 79–150.
Uyama, H. see Kobayashi, S.: Vol. 121, pp. 1–30. Uyama, Y: Surface Modification of Polymers by Grafting. Vol. 137, pp. 1–40.

Usuki, A., Hasegawa, N. and *Kato, M.*: Polymer-Clay Nanocomposites. Vol. 179, pp. 135–195.

Varma, I. K. see Albertsson, A.-C.: Vol. 157, pp. 99–138.

Vasilevskaya, V. see Khokhlov, A.: Vol. 109, pp. 121–172.

Vaskova, V. see Hunkeler, D.: Vol.: 112, pp. 115–134.

Verdugo, P.: Polymer Gel Phase Transition in Condensation-Decondensation of Secretory Products. Vol. 110, pp. 145–156.

Vettegren, V. I. see Bronnikov, S. V.: Vol. 125, pp. 103–146.

Vilgis, T. A. see Holm, C.: Vol. 166, pp. 67–111.

Viovy, J.-L. and *Lesec, J.*: Separation of Macromolecules in Gels: Permeation Chromatography and Electrophoresis. Vol. 114, pp. 1–42.

Vlahos, C. see Hadjichristidis, N.: Vol. 142, pp. 71–128.

Voigt, I. see Spange, S.: Vol. 165, pp. 43–78.

Volk, N., Vollmer, D., Schmidt, M., Oppermann, W. and *Huber, K.*: Conformation and Phase Diagrams of Flexible Polyelectrolytes. Vol. 166, pp. 29–65.

Volksen, W.: Condensation Polyimides: Synthesis, Solution Behavior, and Imidization Characteristics. Vol. 117, pp. 111–164.

Volksen, W. see Hedrick, J. L.: Vol. 141, pp. 1–44.

Volksen, W. see Hedrick, J. L.: Vol. 147, pp. 61–112.

Vollmer, D. see Volk, N.: Vol. 166, pp. 29–65.

Voskerician, G. and *Weder, C.*: Electronic Properties of PAEs. Vol. 177, pp. 209–248.

Wagener, K. B. see Baughman, T. W.: Vol 176, pp. 1–42.

Wake, M. C. see Thomson, R. C.: Vol. 122, pp. 245–274.

Wandrey, C., Hernández-Barajas, J. and *Hunkeler, D.*: Diallyldimethylammonium Chloride and its Polymers. Vol. 145, pp. 123–182.

Wang, K. L. see Cussler, E. L.: Vol. 110, pp. 67–80.

Wang, S.-Q.: Molecular Transitions and Dynamics at Polymer/Wall Interfaces: Origins of Flow Instabilities and Wall Slip. Vol. 138, pp. 227–276.

Wang, S.-Q. see Bhargava, R.: Vol. 163, pp. 137–191.

Wang, T. G. see Prokop, A.: Vol. 136, pp. 1–52; 53–74.

Wang, X. see Lin, T.-C.: Vol. 161, pp. 157–193.

Webster, O. W.: Group Transfer Polymerization: Mechanism and Comparison with Other Methods of Controlled Polymerization of Acrylic Monomers. Vol. 167, pp. 1–34.

Weder, C. see Voskerician, G.: Vol. 177, pp. 209–248.

Whitesell, R. R. see Prokop, A.: Vol. 136, pp. 53–74.

Williams, R. J. J., Rozenberg, B. A. and *Pascault, J.-P.*: Reaction Induced Phase Separation in Modified Thermosetting Polymers. Vol. 128, pp. 95–156.

Winkler, R. G. see Holm, C.: Vol. 166, pp. 67–111.

Winter, H. H. and *Mours, M.*: Rheology of Polymers Near Liquid-Solid Transitions. Vol. 134, pp. 165–234.

Wittmeyer, P. see Bohrisch, J.: Vol. 165, pp. 1–41.

Wu, C.: Laser Light Scattering Characterization of Special Intractable Macromolecules in Solution. Vol 137, pp. 103–134.

Wunderlich, B. see Sumpter, B. G.: Vol. 116, pp. 27–72.

Xiang, M. see Jiang, M.: Vol. 146, pp. 121–194.

Xie, T. Y. see Hunkeler, D.: Vol. 112, pp. 115–134.

Xu, Z., Hadjichristidis, N., Fetters, L. J. and *Mays, J. W.*: Structure/Chain-Flexibility Relationships of Polymers. Vol. 120, pp. 1–50.

Yagci, Y. and *Endo, T.*: N-Benzyl and N-Alkoxy Pyridium Salts as Thermal and Photochemical Initiators for Cationic Polymerization. Vol. 127, pp. 59–86.

Yamaguchi, I. see Yamamoto, T.: Vol. 177, pp. 181–208.

Yamamoto, T., Yamaguchi, I. and *Yasuda, T.*: PAEs with Heteroaromatic Rings. Vol. 177, pp. 181–208.

Yamaoka, H.: Polymer Materials for Fusion Reactors. Vol. 105, pp. 117–144.

Yannas, I. V.: Tissue Regeneration Templates Based on Collagen-Glycosaminoglycan Copolymers. Vol. 122, pp. 219–244.

Yang, J. S. see Jo, W. H.: Vol. 156, pp. 1–52.

Yasuda, H. and *Ihara, E.*: Rare Earth Metal-Initiated Living Polymerizations of Polar and Nonpolar Monomers. Vol. 133, pp. 53–102.

Yasuda, T. see Yamamoto, T.: Vol. 177, pp. 181–208.

Yaszemski, M. J. see Thomson, R. C.: Vol. 122, pp. 245–274.

Yoo, T. see Quirk, R. P.: Vol. 153, pp. 67–162.

Yoon, D. Y. see Hedrick, J. L.: Vol. 141, pp. 1–44.

Yoshida, H. and *Ichikawa, T.*: Electron Spin Studies of Free Radicals in Irradiated Polymers. Vol. 105, pp. 3–36.

Zhang, H. see Rühe, J.: Vol. 165, pp. 79–150.

Zhang, Y.: Synchrotron Radiation Direct Photo Etching of Polymers. Vol. 168, pp. 291–340.

Zheng, J. and *Swager, T. M.*: Poly(arylene ethynylene)s in Chemosensing and Biosensing. Vol. 177, pp. 151–177.

Zhou, H. see Jiang, M.: Vol. 146, pp. 121–194.

Zubov, V. P., Ivanov, A. E. and *Saburov, V. V.*: Polymer-Coated Adsorbents for the Separation of Biopolymers and Particles. Vol. 104, pp. 135–176.

Subject Index

Acetoxy silanes   6
ACPs, sulfonated   88
Acrylic acid   138
Acrylic resin-clay
   nanocomposites   138
AFM, silicate platelets   42
Alkylsulfonation   109
Aluminum   190
12-Aminododecanoic acid   141, 148
Arylsulfonation   111
Arylsulfonic acids, desulfonation   104

Bentonite   36
Bischlorosilanes   14
Bis-hydroxy(tetramethyl-*p*-silphenylene
   siloxane)   4, 7
Bis-silanol   4
– hydrosilylation   14
Butadiene, carboxy-terminated   33
ε-Caprolactam   137, 141

Carbosiloxanes, linear   20
CEC   49
Clay, NMR organoclay layer
   separation   43
Clay dispersion   45
Clay nanocomposites   135
Clay slurry, compounding   153
Composites   30
Condensation polymers   83
– aromatic   87
Copoly(carbosiloxane)s, fluorinated   20
Covulcanization   139
Crack propagation   33
CTBNs   33
Cyanate ester nanocomposites   59
Cycloreversion   2

DABDT   101

DDS   46
1,10-Decanedicarboxylic acid   162
DETDA   41
DGEBA   40, 46
Diacetamidosilanes   17
Diacetylene   9
Diamines, sulfonated   97
4,4′-Diamino-2,2′-diphenylsulfonic
   acid   98
1,10-Diaminodecane   162
α,ω-Dienes   19
Dihydrosiloxanes   1
α,ω-Dihydrosiloxanes   19
Disilanol diaminosilane
   polycondensation   7
DMAC   140
DMTA   57
DSC   6

EPDM-clay   189
Epoxy fiber composites   35
Epoxy resins, layered silicate
   nanocomposites   32
Epoxy thermosets   33
EPR   183
Ethyl acrylate   138
Ethylene propylene diene rubber
   (EPDM)   189
Ethylene propylene rubber
   (EPR)   183
Ethylenethiourea   190
Exfoliation   46
– cure   53

F/silicone homopolymers, hybrid   17
FASIL   12
Fiber, epoxy matrix   35
Fillers   31
Fluoroether   15

Fluorosilicones   13
Fracture   63
Fuel cells, proton-exchanging
       membranes   83, 127

Gallery (interlayer)   37
Glycine   154
Green nanocomposites   192

Heat resistance   9
Hectorite   154
Hexahydrophthalic anhydride
       (HHPA)   61
HMDA   154
Hybrid copolymers, fluorinated   15
Hybrid intercalation   135
Hybrid silicones   1
Hydrosilylation   1

Intercalation   44
Izod impact test   169

Jeffamines   45, 51, 55

Karstedt catalyst   19

Laminates, 3D   35

MA2   171
Maleic anhydride   173
Matrix ductility/toughness   35
Melt flow index   180
Membranes, gas separation   22
– perfluorinated   87
– proton-exchanging
       electrolyte   83, 127
2-Mercaptobenzothiazole   190
Mica, synthetic   154
Microcomposite, layered silicates   32
Molecular weight   7
Monomer intercalation   140
Montmorillonite   29, 30, 36, 41, 140
MPDA   40, 50

Nafion membrane   87
Nanocomposites, silicate,
       epoxy-layered   29
Nanopowders   32
NBR-clay nanocomposite   139
NCC   142

NCHs   142
– alignment of silicate layers   156
– crystal structure   155
– gas barrier characteristics   150
– properties   148
NMR, silicate platelets   43
Nylon-clay nanocomposites   135
Nylon 6, clay gallery   141
Nylon 6 crystals, alignment   159
Nylon 6-clay, flame resistance   163
Nylon 11   162
Nylon 66-clay   162
Nylon 1012   162

Organoclay layer separation,
       NMR   43
Organopolysiloxanes,
       perfluoroalkylene   11
Organosilicate exfoliation,
       cure temperature   53
Oxidation stability   7

Paraffin-siloxanes   3
PBI, alkylsulfonated   109
– phosphoric acid   121
– propylsulfonated   113
– proton conductivity   124
PBTs   101
PCPQ   126
PDMS   5
PEEK, sulfonation   88, 103
PEMFCs   102, 120
Perfluorinated polymers   86
Perfluorocyclobutane   13
Perfluoroelastomer, liquid   23
Perfluoroether   23
PES, sulfonated   127
Phyllosilicates   37
Plant oil-clay nanocomposites   192
Platinum   19
PMMA, thermoplastic matrix   74
PO1015   171
Poly(arylenevinylenesiloxanes)   22
Poly(benzamidazoles) (PBI)   109
Poly(imidesiloxanes) (PI/PS)   21
Poly(naphthylimides)   99, 100
Poly(p-phenylene), sulfonated   94
– sulfonate-substututes   118
Poly(p-phenylene ethynylene)   119
Poly(p-phenylene terephthalamide)   112

Poly(*p*-phenylene terephthalamido-*N*-propylsulfonate) 109
Poly(phenylene sulfide), sulfonated 104
Poly(phthalazinone ether ketones), sulfonated 95
Poly(siloxylene-ethylene-phenylene-ethylene) 9
Poly(styrenesulfonic acid) 86
Poly(tetramethyl-*m*-silphenylene siloxane) 8
Poly(thiophenylene sulfones), sulfonated 96
Polyamic acid 140
Polyamides 136
– sulfonated 97
Polyazoles, sulfonated 100
Polybenzobisthiazoles 101
Polybenzoxazoles, sulfonated 102
Polycarbosiloxanes 1, 19
Polydimethylsiloxane 20
Polyelectrolytes 83, 95
Polyether polyol 74
Polyethylene-clay, gas permeability 182
Polyhydrosilylation 1, 20, 22
Polyimide 140
Polylactic acid 192
Polynaphthalenecarboximides 98
Polyolefin diol 164
Polyolefin-clay 163, 190
Polyolefins 135, 189
Poly-1,3,4-oxadiazole 125
Polyperyleneimide, sulfonated 93
Polyphenylquinoxaline 125
Polypropylene 136, 140, 173
Polypropylene-clay composite 164, 166
Polypropylene-talc 166
Polyquinoline, pyrrole-containing 126
Polysilalkylene siloxanes 1, 2, 23
– fluorinated 18
Polysilarylene siloxanes 1
Polysiloxane, fluorinated 1
Polysiloxane/polyimide block 22
POLYTEL H 164
Polytrifluoropropyl methyl siloxane 12
PP 164, 191
PP(MA2) 170
PPBP 103
PPCN 165

PPTA 112
PPTC-5 171
Proton-exchanging membranes 83
Pt-DVTMDS 19
PTFPMS 18

Resin infusion, film 36, 71
Reversion, resistance 14

Saponite 154
SAXD, silicate platelets 40
Sealant materials 10
Self-passivation 163
SEM, silicate platelets 42
SIFEL perfluoro elastomer 23
Silalkylene-siloxanes, fluorinated 17
Silarylene-siloxane copolymers 1, 6
Silicate nanocomposites, epoxy-layered 29
– morphology, TEM/WAXD 39
Silicate platelets, delaminated 38
– exfoliated 38, 63
– intercalated 38, 63
Silicone rubbers, vulcanized 23
Silicones 1
– fluorinated 10
– pumpable fluorinated 8
Siloxane elastomers, thermoplastic 22
Siloxanes 1
Silylfluoroaromatic homopolymers 11
Smectites 36
– charge density 49
Solvents, resistance 14
Speier catalyst 19
S-PPBP 103
Storage moduli 177
Swelling tests 187

Tactoids 38
TEM, silicate platelets 40
Tensile test 170
Tetramethylthiuram monosulfide 190
TGA 6, 66
Thermal relaxation 58
Thermal stability 7, 20
Thermosets, epoxy 33
– particle toughening 33
– rubber-toughened 33
– thermoplastic toughening 34
Thermosetting nanocomposites, cure 55

TMPS-DMS   7
Trienes, hydrosilylation   23
Triglyceride oils   192

Ureidosilanes   6

Vinyl ethers, perfluorinated   86
Vulcanization accelerators   190

WAXD, silicate platelets   39
Wet winding   71

Zinc dimethyldithiocarbamate   190